The European Union as Crisis Manager

The European Union is increasingly being asked to manage crises inside and outside the Union. From terrorist attacks to financial crises, and natural disasters to international conflicts, many crises today generate pressures to collaborate across geographical and functional boundaries. What capacities does the EU have to manage such crises? Why and how have these capacities evolved? How do they work and are they effective? This book offers a holistic perspective on EU crisis management. It defines the crisis concept broadly and examines EU capacities across policy sectors, institutions, and agencies. The authors describe the full range of EU crisis management capacities that can be used for internal and external crises. Using an institutionalization perspective, they explain how these different capacities evolved and have become institutionalized. This highly accessible volume illuminates a rarely examined and increasingly important area of European cooperation.

ARJEN BOIN is Professor of Public Governance and Crisis Management in the School of Governance at Utrecht University, the Netherlands, and an Adjunct Professor in the Public Administration Institute of Louisiana State University. He is also Managing Partner at Crisisplan.

MAGNUS EKENGREN is an Associate Professor at the Swedish National Defence College, and Director of the College's Programme for European Security Research (EUROSEC).

MARK RHINARD is Senior Research Fellow at the Swedish Institute of International Affairs and an Associate Professor at Stockholm University.

The European Union as Crisis Manager

Patterns and Prospects

ARJEN BOIN

Utrecht University & Louisiana State University

MAGNUS EKENGREN

Swedish National Defence College

MARK RHINARD

Swedish Institute of International Affairs & Stockholm University

CAMBRIDGE
UNIVERSITY PRESS

CAMBRIDGE
UNIVERSITY PRESS

University Printing House, Cambridge CB2 8BS, United Kingdom

Published in the United States of America by Cambridge University Press, New York

Cambridge University Press is part of the University of Cambridge.

It furthers the University's mission by disseminating knowledge in the pursuit of education, learning and research at the highest international levels of excellence.

www.cambridge.org
Information on this title: www.cambridge.org/9781107035799

First published 2013

Printed and bound in the United Kingdom by Bell and Bain Ltd

A catalog record for this publication is available from the British Library

Library of Congress Cataloguing in Publication data
Boin, Arjen.
The European Union as crisis manager : patterns and prospects / Arjen Boin, Utrecht University & Louisiana State University, Magnus EKengren, Swedish National Defence College, Mark Rhinard, Swedish Institute of International Affairs & Stockholm University.
pages cm
Includes index.
ISBN 978-1-107-03579-9 (hb.)
1. Emergency management – European Union countries. 2. Crisis management – European Union countries. I. Ekengren, Magnus. II. Rhinard, Mark, 1973– III. Title.
HV551.5.E85B65 2013
363.34′8094 – dc23 2013017283

ISBN 978-1-107-03579-9 Hardback
ISBN 978-1-107-68028-9 Paperback

Additional resources for this publication at www.cambridge.org/boin

Contents

Preface

Why write a book about the European Union as a crisis manager? The response to safety and security threats falls in the traditional domain of the nation state. What role could the EU possibly play when it comes to dealing with floods, electricity breakdowns, epidemics, or terrorist attacks?

Our search for an answer to this question began some ten years ago. In those post-9/11 days, two of the authors met over coffee at Leiden University's Department of Public Administration. One (a crisis management researcher) had been studying how national governments were preparing for the possibility of a smallpox attack (a major worry in those days). He had understood that small countries like the Netherlands could never cope with such an event and would rely on cooperation with its neighbors. Surely, he asked, the European Union would play a dominant role in the response to a smallpox outbreak?

The other author (an EU researcher) found this a silly question. He explained that the EU was never intended or designed to "manage crises." True, the EU did send military and civilian missions to foreign hot spots and there was financial support for humanitarian aid in disaster-stricken countries. But the EU did not manage crises or disasters on the European continent. Yet the question lingered: what role *could* the EU play in response to a large-scale crisis or disaster?

Serendipity brought the third author into the team. A security studies scholar and former diplomat, he had become convinced that traditional security approaches could not account for the EU's expanding crisis management capacities in the foreign domain. A chance meeting in Stockholm gave rise to a joint effort, which has culminated in this book.

Our team combines insights from EU studies, security theory, public administration, and crisis management research. Different intellectual backgrounds, methods of conducting research, ways of writing, and styles of working can prove quite frustrating (we have our stories). But we found that different viewpoints also nurture fresh perspectives and lead to new insights.

This has been a fun project. We interviewed scores of policymakers and experts, presented our insights to practitioners and academics, and learned many new things about the EU and its many institutions and policy sectors that harbor crisis management capacity. We discovered that much more was happening than we had imagined. We also learned that it is not easy to align the interests and resources of EU member states with the crisis management capacities spread across the EU's administrative apparatus. Yet, we came away convinced that the EU has a role to play in responding to a host of new threats.

Who might benefit from these findings? We think three academic communities will find this study of interest. Security scholars should find the empirics refreshing and theoretically challenging. Our research helps to understand the EU's shifting security "identity" – a preoccupying concern of security scholars studying the European Union. EU scholars will learn about pockets of ongoing cooperation in an era when European integration appears to be stalling. Public administration scholars, especially those with an interest in crisis management, will be surprised to learn about the emerging architecture of a crisis management system at the European level.

This research project has been a long and rewarding journey. Along the road, we were fortunate to receive a helping hand from many friends and colleagues. They offered inspiration and direction and saved us from critical errors – those remaining in this manuscript are due to our own shortcomings.

We are especially indebted to Bengt Sundelius, who believed in this project from the very beginning and commented on more manuscripts than we can remember. This book would not have been written without his encouragement, guidance, and support.

We have benefited from the invaluable advice and assistance of academic colleagues across the world, including Chris Ansell, Alyson Bailes, Tom Birkland, Raphael Bossong, Hugo Brady, Malena Britz, Emery Brusset, Fredrik Bynander, Amy Donahue, Simon Duke, Kjell Engelbrekt, Ana Mar Fernandez Pasarin, Rhona Flin, Tuomas Forsberg, Martijn Groenleer, Helga Haftendorn, Paul 't Hart, Timo Hellenberg, Regina Heller, Margaret Hermann, Adrian Hyde-Price, Knud Erik Jørgensen, Emil Kirchner, Patrick Lagadec, Todd LaPorte, Kelley

Lee, Javier Lezaun, Gustav Lindström, Martin Lodge, Ian Manners, Allan McConnell, Antonio Missiroli, Jörg Monar, Mirko Noordegraaf, Hanna Ojanen, Charles Parker, Salvador Parrado, Boris Porfiriev, Iztok Prezelj, Sebastiaan Princen, Emery Roe, Adam Rose, Helene Sjursen, Jim Sperling, Carl-Einar Stålvant, Eric Stern, Kai Wegrich, and Cecile Wendling.

Special thanks go to the practitioners who gave of their time during interviews and provided us with documents and key insights. We are especially grateful to the following practitioners for their help: Maria Åsenius, Emanuela Bellan, Hans Das, Jan van Elst, Johnny Engell-Hansen, Vera Ferreira, Florika Fink-Hooijer, Kristalina Georgieva, Henk Geveke, Corsmas Goemans, Ana Gomes, Roberto Gonella, Michalis Ketselidis, Thomas de Lannoy, Katell Le Goulven, Cecilia Malmström, Agostino Miozzo, Gerrard Quille, David Spence, Anders Tegnell, and Margot Wallström.

The Swedish Civil Contingencies Agency has been instrumental to this research program. It has generously funded our project without imposing the type of micro-managing oversight that has become so commonplace in the world of academic research. We are grateful to Helena Lindberg, the General-Director of the Swedish Civil Contingencies Agency, who supported our research and participated in a number of panels and conferences. Anna Halldén, our research contact point at the agency, offered valuable advice and lots of enthusiasm. We salute the agency for its commitment to academic research and hope it will serve as an example to other government agencies.

We have been privileged to enjoy a fruitful partnership with the European Policy Centre, including its director, Hans Martens, and its previous research director, Antonio Missiroli, as our research program unfolded. That partnership proved essential not only in facilitating the collection of data but also in disseminating our results to the EU policy community.

We could not have carried out this project without the help of our research group, which served as the engine that kept the program moving. The invisible force behind our team is Sara Myrdal, who provided us with many case studies, policy briefs, and sharp commentaries. More importantly, her spirited company and astute insights have greatly contributed to this project. In the early years, Sanneke Kuipers, Nina Wilhelmson, and Monica Svantesson did much of the empirical "heavy lifting" that provided a solid basis for ongoing analysis. Mette Bakken, Lavinia Cadar, Adriana Fagarasan, Emma Hallencreutz, Simon Hollis, Kristian Julin, Kristin Ljungkvist, Christer Magnusson, and Cecilia Mobach provided detailed input and "back office" support. Åsa Fritzon and Erik Brattberg have been constant sources of research support, ranging from debating ideas to providing editorial support and from organizing events to collecting citations. Marlot Kuiper spent countless hours putting the reference list together. Charlotte Boin and Caroline Richards provided copyediting support during the final phase of the project.

We wish to thank John Haslam of Cambridge University Press for his trust and support of this book project. We extend our gratitude to the two anonymous reviewers who provided us with detailed criticism and corrected factual errors and helped us sharpen the argument of the book. We thank Carrie Parkinson for shepherding the manuscript through the production process.

Finally, we have some individual debts to acknowledge. Arjen Boin wishes to express his gratitude to Jim Richardson, chair of Louisiana State University's Public Administration Institute, for creating an excellent research environment for this project. Thanks go out to colleagues at the Utrecht School of Governance, especially Mark Bovens and Sanneke Kuipers, for allowing time to write this book. Werner Overdijk, managing partner at Crisisplan, deserves praise for his endless patience and understanding. Deepest gratitude is reserved for Mary, Charlotte, and Désanne, who lovingly allowed this project to move forward (even when other issues demanded and deserved attention).

Magnus Ekengren is very thankful to the Swedish National Defence College and all the colleagues and friends at the Department of Security, Strategy and Leadership for encouragement, writing time, and inspiration – all those things that characterize a productive research milieu. A special thank you goes to the members of the European Security Research Programme (EUROSEC). He owes the warmest thanks to Paula, whose love and support have made life bright despite hard work and European crises.

Mark Rhinard extends his thanks to the Swedish Institute of International Affairs and to Stockholm University, for providing research time that enabled the completion of this book. He dedicates his contribution herein to Karin for lovingly creating the environment for a perfect work–life balance and to Tim, who was born as this book came to fruition and who has turned out to be anything but a crisis.

1 *The EU as crisis manager*

A new role for the Union

The surprising emergence of the EU as crisis manager

There are few reasons to expect the European Union (EU) to play a role in the management of crises and disasters. The response to such events has traditionally been the remit of national governments. What could the EU – often depicted as a bureaucratic talk shop – possibly add to the efforts of national and local governments?

Quite a bit, as this book reveals. The member states have invested the EU with a significant amount of what we refer to as "crisis management capacity." While often reluctant to transfer more authority to Brussels, member states have shown a sustained willingness to enhance the EU's crisis management capacities. After a large-scale crisis or disaster, member states routinely call for additional EU capacities to coordinate, link, or integrate their response capacities. Few European Council meetings conclude without some call for more crisis cooperation.

The EU has indeed become more visible as a crisis manager in recent years. Consider the following examples:

- In January 2010, a massive earthquake struck Haiti. The EU coordinated the humanitarian response of its member states, sent a police force of 200 Europeans, and created a relief fund for the devastated island.
- In the early months of 2011, popular revolts broke out across northern Africa and the Middle East. The EU sent its High Representative, Catherine Ashton, to newly liberated countries to assess how the EU could help their democratic development.[1] The EU imposed an arms embargo on Libya and discussed the imposition of a no-flight zone. Meanwhile, the southern member states appealed to the EU for a coordinated response to the feared exodus of young Arabs seeking a better future on the European continent.

- In the spring of 2011, a vicious *E. coli* (EHEC) epidemic in Germany caused the deaths of over 40 people. After Germany informed the European Commission through the EU's Early Warning and Response System, the Commission's DG Sanco (Directorate General for Health & Consumers) took the lead in coordinating an EU-wide investigation and control measures.[2]
- In 2012, the euro's future was cast in doubt and several member states were teetering on the brink of financial disaster. In response, the European Central Bank took a series of unprecedented actions to maintain stability in the Eurozone. Its response to the financial crisis fit a wider pattern of the EU breaking traditions, initiating new policies, imposing new rules, and creating institutions to deal with this and other deep crises.

This book shows that the EU has in place a substantial and growing capacity to initiate and coordinate a shared response to many different threats, crises, and disasters. The EU harbors a wide variety of policies, mechanisms, and institutions that can, and do, facilitate a coordinated response of member states. When member states want to work together to deal with a crisis at home or abroad, they can use EU venues and mechanisms to facilitate a joint response.

These coordination mechanisms fail at times, it is true. When the EU stumbles in response to crisis, often due to the constraints member states imposed on it, these same member states stand ready to criticize the Union as slow and ineffective. But while the EU is far from perfect, it may well be the best option available to its member states. As more and new crises appear on the horizon, member states will have to increase mutual cooperation. That will increase pressure on the EU to act faster and perform better.

The adoption of the Lisbon Treaty has brought some intriguing developments that may have critical consequences for the EU's crisis management role. The High Representative now commands the European External Action Service, which should further enhance Europe's role abroad. The Lisbon Treaty contains the so-called Solidarity Clause, which implores member states to assist each other in case of a terrorist attack or natural disaster. The EU now has a Commissioner dedicated to Home Affairs, a budding Internal Security Strategy, and growing disaster management capacities.

All efforts to enhance the EU's role in the domain of crisis management inevitably encounter a deep-seated problem: there is very little agreement on how crisis and security management efforts – traditionally the responsibility of member states – can be effectively and legitimately organized at the European level. Despite the gradual accumulation of crisis-relevant tools, instruments, and decision venues, national leaders have avoided taking a stance on, or clarifying, the EU's role in managing crises when given the opportunity to do so in major treaty revisions. Despite their regular calls for more cooperation, member states introduced a provision in the Lisbon Treaty that "national security remains the sole responsibility of each Member State" (Art. 4).[3]

There is, in other words, a deep tension between the often pronounced ambition to make Europe (and the world) a safer place and the entrenched principle of subsidiarity that has guided the historic development of the Union.[4] The result, as we show in this book, is an unclear division of competences between the national and European levels.

In this book we describe what kinds of crisis and disaster management capacities the Union has and where these are located in its institutional architecture. We explain why and how the EU has developed these capacities. We explore how the EU can further develop its capacities to deal with the crises that member states will face in the future.

We begin this chapter by exploring the world of crisis and argue that more intensive cooperation between nation states is needed. We provide a brief overview of recent developments in the EU that signify the emergence of a safety and security area in the Union. We then elaborate our research questions, introduce our theoretical framework, and present the outline of this book.

A new world of crisis: modern vulnerabilities, urgent threats, and impossible challenges

We live in what Ulrich Beck calls "the global risk society" (Beck 1992, 2008; cf. Perrow 2007). Climate change, tilting geopolitics, new forms of terrorism, and financial upheaval – these are but a few of the "known unknowns" that challenge the Western way of living. These security threats originate from many sources, cross political and functional boundaries with ease, and have the potential to affect a wide variety of critical infrastructures (LaPorte 2007). Their potential for devastation is immense (OECD 2003; Lomborg 2004; Posner 2004; Missiroli 2006; Perrow 2007; Alemanno 2011; Helsloot *et al.* 2012).

Societies have always faced a wide variety of threats to the safety and well-being of their citizens.[5] War, famine, severe weather, earthquakes, floods, and epidemics have brought societies to the brink of collapse (Diamond 2005). It is no wonder, then, that crisis management – the set of activities aimed at minimizing the impact of a crisis – has traditionally been recognized as a core function of the state (Rosenthal 1980; Schmitt 1985). In times of crisis, people look to their leaders (elected or not) to protect them from the looming consequences. A failure to do so inevitably undermines the legitimacy of the governing elites and their institutions (Boin *et al.* 2005).

Research shows that governments typically find crisis management no easy task (Rosenthal, Charles, and 't Hart 1989; Schneider 1995; Rosenthal, Boin, and Comfort 2001; Kettl 2004; Boin *et al.* 2005; Drennan and McConnell 2007; Sylves 2008). Planning and preparing for the unknown and unexpected – the proverbial black swans – tests the limits of what bureaucracies are designed to do (Clarke 1999; Taleb 2007).[6] During a crisis, governing elites wrestle with the challenges of sense-making (creating a shared picture of the situation), making critical decisions with limited or incorrect information, coordinating a large-scale response with many actors, and communicating effectively with the public and stakeholders. There are many things that can go wrong during a crisis (Boin and 't Hart 2010).

And these crisis governance challenges are not getting any easier. The critical importance of creating effective crisis management structures and practices is cast in a new light by the changing nature of threats to modern societies. Two developments, in particular, conspire to raise the stakes.

First, threat agents are changing. New technologies fuel progress, but they also "bite back" (think of infrastructural breakdowns, cyber warfare, and engineered viruses).[7] Climate change may bring natural destruction and altered living conditions (Brauch *et al.* 2008). The rise of artificial intelligence, the cheapening of DNA-modification techniques, and the preparations for cyber warfare can create unforeseen and perhaps unimaginable crises (Dror 2001; Krepinevich 2009). Traditional threats – epidemics, terrorism, technological failure, fires, floods, riots – will still be with us, of course. But they may take new and unforeseen shapes (Rosenthal, Boin, and Comfort 2001).

A second, more insidious development compounds the effects of the first. As modern societies relentlessly weave their critical infrastructures into a global fabric, new vulnerabilities of many kinds are being created. The modern, open society has helped to generate gains in prosperity that seemed unimaginable just a few decades ago (the EU is a textbook example). But the same drivers of prosperity also enable threat agents to infiltrate these societies through the mechanisms of integration. Terrorists, technical failures, financial crises, and health epidemics exploit our modern systems, which allow them to unleash disruption and destruction (cf. Turner 1978).

Two recent crises – the Icelandic ash cloud and the breakdown of the financial system – underline the global reach and escalatory potential of modern crises and demonstrate the serious challenges these crises pose to public authorities (Boin *et al.* 2005; Ansell, Boin, and Keller 2010; Alemanno 2011). Prevention is nearly impossible, as crises can originate in faraway domains and travel across transboundary systems that cannot be shut down without incurring huge costs.

National governments are ill equipped to address these complex challenges (Boin 2005; Lagadec 2009).[8] When national bureaucracies have to work across policy boundaries or geographical borders (with other national bureaucracies) to face urgent threats, paralysis looms large (Rhinard 2009).[9] In light of future crises, they will need a way to overcome these barriers and work together. The EU is increasingly the place where such attempts are being made.

Looking for EU capacities: three crisis domains

This book identifies capacities that the EU can use to facilitate a coordinated response of its member states to large-scale crises and disasters, wherever these may occur. Few crisis management scholars have studied the EU. Students of the EU, in turn, have traditionally studied particular areas of crisis management such as the EU's Common Security and Defence Policy (CSDP) (Howorth 2007; Jones 2007), the area of Justice and Home Affairs (Mitsilegas, Monar, and Rees 2003; Lavenex and Wallace 2005), health security (Sundelius and Grönvall 2004), terrorism (Bossong 2008), critical infrastructure (Pursiainen 2009), or intelligence cooperation (Walsh 2006). We employ a comprehensive approach in our study of the EU's crisis management capacities.

Before we explain how we identify EU crisis management capacities, we should say something about our choice of terminology here. More specifically, we should explain our use and interpretation of the word *crisis*.

The term *crisis* has enjoyed popular use across academic disciplines as well as in everyday conversation. In these discourses, the term is used in different and varying ways. Virtually all definitions refer to some threat that must be urgently averted or addressed in order to avoid dire consequences. But threat agents and objects of threat can vary widely, depending on what definition is used (Boin 2005). It is, thus, crucial to specify what "crisis" – a central concept in our book – means.

Let us begin by adopting a very generic definition of crisis, in terms of "a perceived threat to the core values or life-sustaining systems of a society that must be urgently addressed under conditions of deep uncertainty."[10] This definition pertains to a wide variety of threats and disasters, which can play out at the local, regional, continental, or global level.

It is important to point out the subjective nature of this crisis definition. In this approach, it is a crisis when political-administrative elites conceive of a situation in terms of a serious threat that requires an urgent response. If politicians and policymakers do not recognize a threat even if many others do (think of climate change or pension deficits), it is an ignored crisis at best. We should, therefore, consider how this generic crisis concept relates to threat perceptions in the community of EU scholars and practitioners.

That is easier said than done. Different policy communities within the EU have different ideas about which threats the EU should face and address; the same can be said for EU scholars who, depending on their object of study, wield different threat conceptions. There are quite a few scholars who use the term *crisis management* to describe what takes place in the EU's policy area of common security and defense (Smith 2003; Howorth 2007). Many speak of "crisis" in reference to the recent financial and economic implosion in Europe (Begg 2010; Baldwin, Gros, and Laeven 2010). Students of the EU's Justice and

Home Affairs domain primarily refer to such perceived threats as the uncontrolled influx of migrants or impending social unrest when they use the term *crisis* (Liddle and Diamond 2010; Johns 2011).

Rather than imposing one encompassing crisis definition on these various communities, we have inductively (based on our reading of the literature and our conversations with many policymakers) identified three "crisis types" currently occupying the attention of EU policymakers (and scholars). In this book, we distinguish between the national crisis (Type I), the external crisis (Type II), and the transboundary crisis (Type III).[11] All three types have elicited a sustained response – in terms of agenda attention and capacity building – from the European Union and its member states. As a result, the EU's capacities are spread over different policy communities. Those capacities are the focus of this book.

The national crisis (Type I). At times, a country may face a disaster that outstrips its capacity to respond. Classic examples are floods (the 1953 flooding of the Dutch province of Zeeland stretched national capacities to the limit), earthquakes (the 1980 earthquake in southern Italy killed thousands), and forest fires (the southern member states regularly experience forest fires that run out of control).

We speak of an overwhelmed state when the available resources are not (or no longer) sufficient to alleviate the suffering of the victimized population. This rarely happens. Most developed nations are prepared to deal with natural disasters, terrorism, explosions, epidemics, collective violence, and even war.

But when the unthinkable happens, member states can call on the EU to coordinate assistance from other states. The Solidarity Clause in the Lisbon Treaty enjoins member states to assist an overwhelmed member state. The EU has in place a Civil Protection Mechanism to facilitate cooperation between member states willing to help out a disaster-struck member state. Over the years, this Civil Protection Mechanism has grown into a network of standby experts, a database of available resources, and an organizational structure (the Monitoring and Information Center) that can coordinate rapid resource sharing in times of need.[12]

It is good to note here what will *not* happen in a Type I scenario. First, the EU cannot "take over" or step in as a crisis manager. The EU has no legal or political authority to do that, nor could it currently perform this task if a member state asked. Second, the EU will not monopolize the support function. Member states are free to invoke bilateral assistance or request help from international organizations such as the Red Cross. The EU, at best, supports national governments in a Type I scenario.

The external crisis (**Type II**). When EU officials or scholars use the term *crisis*, there is a good chance they are referring to an international conflict, a large-scale disaster, or a failed state outside EU territory. Archetypes include the Balkan Wars, the Turkey and Haiti earthquakes, the civil war in the Congo, and the revolution in Libya (for which an EU "crisis management" mission was approved but never deployed). Using the term *crisis* in this way became dominant in the 1990s when the EU sought to develop a Common Foreign and Security Policy and was looking for a niche in the international arena to assert its presence (Cottey 2007).

The combination of joint military units (created as a result of the EU's Common Security and Defence Policy), humanitarian funding mechanisms (handled by DG ECHO), and the Civil Protection Mechanism (described above) has provided the EU with capacity to act in foreign hot spots. The EU has, for instance, assisted in disaster zones in Turkey, Morocco, and Asia, offered support to US authorities after Hurricane Katrina, and assisted flood-stricken towns in Central Europe and Algeria. The EU now routinely deploys civil protection experts, police authorities, judicial advisors, and civil administration officials to help stabilize postconflict regions anywhere in the world.

However modest these operations may have been, they mark the arrival of the EU as an international crisis manager. The EU's ambitions in this regard find their most astute pronunciation in the 2003 European Security Strategy (ESS). The ESS identifies crisis events deserving EU attention, ranging from failed states to global warming, from terrorism to natural disasters. It takes a comprehensive view of crisis, explicitly linking internal and external threats, civilian and military capacities, and natural and man-made disasters.

External crises and disasters pose complex challenges. One challenge is of a diplomatic nature: the EU and its member states must operate outside their domain of sovereignty and therefore need to create legitimacy for any intervention it seeks to initiate. It must work with affected countries and international organizations to gain access. This challenge is compounded by the complexities of international operations in crisis areas. Getting resources to disaster or war-stricken areas is no easy task.[13] All this requires negotiation of a third challenge: the EU needs to align the national interests of its member states and facilitate a coordinated and well-resourced response.

While the EU still finds it hard to operate with one voice on the world stage, it has become quite successful, in a somewhat stealthy way, in putting together joint missions to faraway hot spots.

The transboundary crisis (Type III) differs from both other types as it plays out at the transnational level, affecting more than one member state at the same time, often with an impact on multiple sectors or systems. We speak of a transboundary crisis when the life-sustaining systems or critical infrastructures of multiple member states are acutely threatened.[14] The crossing of boundaries thus sets this class of adversity apart from its more localized brethren (the Type I and II crises).

We should, for example, think of breakdowns of the Internet, electricity production, gas supply, or air control systems. Freak weather events, possibly as a result of climate change, can cause problems that affect entire regions.[15] We can think of virulent forms of pandemic influenza or large-scale attacks with chemical or biological weapons that incapacitate large swaths of the European populations and affect schools, businesses, and public services.

A transboundary crisis has no – or at least not one – Ground Zero. This crisis type is extremely complex: it expands in severity and transforms as it moves and crosses infrastructures that themselves are complex and difficult to understand. It can easily create an authority vacuum since it is not clear who "owns" the crisis and who must deal with it. This authority vacuum allows familiar tensions to play up and feed off each other: nation states versus international organizations; central authorities versus local first responders; public organizations versus private interests; state concerns versus citizen fears.

The SARS epidemic is a good example. Its characteristics or causes were ill understood when the disease rapidly proliferated from the southern provinces of China to at least 37 countries. While the number of known fatalities was relatively low (less than a thousand), its effects were widespread. The disease affected the travel and tourism industry, tarnished the reputation of Toronto (as the disease mysteriously lingered in the city), and forced Chinese authorities to revamp their crisis management system.

Europe has faced its share of transboundary crises, including the Chernobyl disaster, the "mad cow" epidemic (BSE), electricity failures, and the Icelandic ash crisis. One of the transboundary types that periodically reappears has to do with illegal immigrants. When, in the 1970s and 1980s, Vietnamese boat refugees and Tamils fled to Europe, they overwhelmed asylum procedures of several member states (adding to the rising antiforeigner sentiments in economically depressed Europe). The refugee flows caused unexpected crises in the member states (Alink, Boin, and 't Hart 2001). This and other refugee crises disclosed Europe's lack of policy coordination and the absence of diplomatic cooperation to stem the flows.

National governments cannot deal with these crises unilaterally, nor can they isolate themselves from these threats. The world – especially Europe – has become too connected and too intertwined to avoid the reach of these crises. It has become nearly impossible to close borders and "decouple" critical infrastructures. Imagine trying to close airports in the Netherlands, while Germany and Belgium keep theirs open. Consider the hoarding of vaccines in one country, when that country's own safety depends on containing a disease in a faraway place in Europe.

Transboundary crises by their very nature require a collective response. A network of local, national, and international actors must be cobbled together. This network has to be adaptable and scalable (Ansell, Boin, and Keller 2010). It has to cross boundaries among units, organizations, sectors, professions, and political jurisdictions; it must be able to validate information, facilitate and communicate decisions, and coordinate the implementation of these decisions.

If there is one domain in which we expect the EU to provide critical assistance (or, in EU speak, to "add value") to national efforts, it is probably this transboundary crisis domain. This book shows that the EU has more capacities in this domain than previously understood. It also reveals that it is the least developed of the three crisis domains.

An institutional framework for analysis

We seek to explain why and how the EU has developed the crisis management capacities it has. How did supranational rules, procedures, venues, and action repertoires emerge and become embedded in each of the EU's crisis domains? How did these structures and processes become legitimate in the eyes of internal and external stakeholders?

We are, in other words, studying the *institutionalization* of the EU's various crisis domains. Institutionalization refers to the process by which a norm, a law, a practice, or an organizational structure takes on institutional characteristics.[16] It is the process through which processes and structures emerge, become embedded in supranational rules, procedures, and patterns of interaction, and gain legitimacy (Selznick 1957; March and Olsen 1984; Suchman 1995).

We use an institutional framework to analyze the three separate policy areas – corresponding with the three crisis types outlined above – with their very different histories, decision-making processes, and institutional characteristics. More precisely, we combine a traditional institutionalist perspective (Selznick 1957; March and Olsen 1984; Finnemore and Sikkink 1998) with explanatory frameworks used to understand the development of EU policy cooperation (Sandholtz 1993; Peterson and Bomberg 1999; Niemann 2006).

EU scholars typically refer to institutionalization as "the process through which a European political space – supranational policy arenas or sites of governance, structured by European rules, procedures, and activities of the EU's organizations – has evolved" (Stone Sweet, Fligstein, and Sandholtz 2001: 3). A high level of institutionalization in this perspective indicates "more Europe," to put it bluntly. As a European political space takes shape, it means that governance increasingly shifts to a supranational level.

Institutionalization can be gauged on a continuum, ranging from lowly institutionalized "intergovernmental" modes of governance (like what takes place in the UN) to fully institutionalized "supranational" modes of governance (more like a federal system with powers over constituent parts). The EU has emerged somewhere in between.

What drives institutionalization in the EU? Stone Sweet, Fligstein, and Sandholtz (2001) argue that over the past decades, a rise in cross-border exchange drove governments, private actors, and civil society representatives to increasingly seek policy solutions at a supranational level. Governments, pressured by private and public actors, but also cognizant of the need for new solutions to new types of public policy problems, agreed to the design and establishment of new arenas for cooperation.

They argue that, over time, ongoing activity and interaction led to those forums becoming "institutionalized," through the accumulation of new rules, norms, procedures, and laws (Sandholtz and Stone Sweet 1998: 7–15). Once supranational rules are in place and new policy spaces have been organized (i.e. institutionalized), those arenas exert influence over the actors that use them. That influence, along with the efforts of EU organizations such as the European Commission, structure the further expansion of cross-border exchange, transnational policy networks, and the EU's policy competences.

This institutionalization approach within EU studies takes somewhat of a linear and almost teleological view of European integration. The process, as depicted in this approach, inevitably leads to more European integration, regardless of the actors' original intentions or the desirability of such outcomes.

Stone Sweet, Fligstein, and Sandholtz acknowledge the functionalist nature of their approach and suggest that the process could, in theory, operate in the opposite direction. However, they are skeptical of "roll back" in the process of institutionalization: "Unraveling institutions, let alone starting again, can be costly or impossible to the extent that actors have invested resources, and adapted their routines, expectations, and relationships with one another, in step with how institutional arrangements have evolved" (Stone Sweet, Fligstein, and Sandholtz 2001: 18).

Understanding institutionalization of the EU's crisis management domains

We assume that institutionalization is neither a random nor a teleological process. We thus look for factors and conditions at different levels of the EU governance system, from the political to the bureaucratic and operational, that combine to drive, impede, or erode the institutionalization of EU crisis management capacities.

In our view, institutionalization resembles an evolutionary process that moves through analytically distinct stages or phases (cf. Finnemore and Sikkink 1998; Boin and Christensen 2008). A capacity can be barely institutionalized, meaning that few people recognize, use, or value it. The same capacity can become deeply institutionalized, meaning it is highly valued, often used, and more or less taken for granted. We take the position that roll back, or at least benign neglect of former initiatives, is possible. Institutionalization may give way to deinstitutionalization. Even deeply institutionalized policy domains may become subjected to institutional erosion.

Drawing together organizational institutionalist approaches and European policy development models, we can identify a pathway of institutionalization (cf. Moravcsik 1998; Peterson and Bomberg 1999; Pierre and Peters 2009; Zeitlin 2011). The underlying drivers of this process are ideas and experiments that take hold when they demonstrate functionality ("added value" to use EU parlance) and gain legitimacy (support from member states). A loss of functionality or legitimacy, in turn, leads to deinstitutionalization.

The institutionalization of EU crisis management capacities: a phase model

In our model, the process of institutionalization starts with *vaguely articulated political ambitions to solve a problem*. That problem is typically a crisis or disaster that, member states might claim, deserves the attention of the EU. The problem is described in sweeping terms; solutions are promised without much specification (Lindberg 1965; Akman and Kassim 2009). Such vaguely articulated ambitions reflect the kinds of "policy-setting" decisions taken by national governments – sometimes at EU summits, sometimes at Council meetings – which set the broad parameters for new European-level action (Peterson and Bomberg 1999).[17] These decisions, while setting the agenda and providing broad legitimacy, rarely include content detail or specify how things should be done. Every intergovernmental summit produces such statements – some prompt feverish amounts of new initiatives and some are never heard of again (Pierson 1996).

Political ambitions can give rise to *an experiment*, typically because a powerful member state or entrepreneurial member of the Commission or Council Secretariat sets the wheel in motion. A little funding and a dedicated bureaucratic unit can bring the experiment to life. First steps may only represent symbolic actions. It can start with a White Paper or a funded research project. Networks of stakeholders may be formed to generate ideas and specific policy options.[18]

When new ideas lead to policies or organizational forms that appear to address a politically salient problem, these actions will be advertised.

If the experiment is perceived to prove its worth and is received favorably, it can reach the next stage of institutionalization.

An experiment will likely face *emerging problems*. When policy meets practice, things rarely occur as originally envisioned. This will produce lessons, both negative and positive (Nye 1987; Checkel 2001; Baumgartner and Jones 2002; Petrov 2010). If positive lessons emerge, and if the experiment promises to address the broadly formulated political aim without much cost, it will be allowed to continue. Then the lessons will be implemented to enhance the policy or organization under development. Experiments that are viewed as suboptimal in addressing problems or used as scapegoats for perceived failures are usually forced to retool, revise, and reinvent its practices (if not its objectives).[19]

Whether an experiment is allowed to continue in the face of negative feedback depends, at least partially, on an organization's *adaptive capacity*. One strategy is to specify and reformulate the policies that gave rise to the experiment to have these policies reflect the capacities under development. This phenomenon is well known in the study of government: existing institutional forms take on new aims and previous tools are used for revised purposes. One can see why such dynamics are common in the EU: having received one-time political approval for experimentation, great effort will be spent on keeping an institutional form alive, even if that means contorting its purposes beyond original intent – to avoid the need for new political approval (Pollack 2006). But an experiment can only lead to institutionalization when it is allowed to become routinized.

Institutionalization crucially depends on efforts to *generate legitimacy* for the capacity under construction (Suchman 1995; Boin 2001). A new type of activity or practice will have to be supported, at least tacitly, by the member states if it is allowed to persist and grow. This means it will have to be "sold" to the member states. Successes must be celebrated, emerging weaknesses quickly addressed. Other stakeholders (media, interest groups, EU institutions) must be kept on board (Pierre and Peters 2009).

Institutionalization is a cycle kept in motion by feedback. Positive feedback – measured in terms of political support (or non-intervention) – maintains and strengthens the institutional level. Negative feedback, which comes in the guise of negative media attention and political discontent (Baumgartner and Jones 1993), has an eroding effect and will require adaptation. When left unaddressed, it can lead to sudden reversals or terminations.

One major source of "feedback" that can change policy trajectories quickly is a crisis (Boin and 't Hart 2000). A terrorist attack, for instance, not only traumatizes a society but also "shocks" the institutionalization process we describe here. National representatives arrive in Brussels clamoring for new solutions. Sweeping aims are formulated and immediate action is announced. Idling experiments are rediscovered and reinvigorated. And so the cycle begins again.

Studying institutionalization: our approach

This book maps a new European policy space that has emerged over the past two decades or so.[20] This space encompasses a large number of policy sectors, crosses traditional EU policy boundaries, is populated by an array of government and societal actors, and includes a variety of institutional venues. Its contours span the internal and external divide that traditionally splits the EU policymaking spectrum. Its policies may exist in institutionally divided units and may be managed through different legislative procedures, but they are all concerned with crisis and disaster management at the European level.

The book tries to understand how, to which degree, and why certain crisis management capacities have become institutionalized at the EU level. In studying institutionalization, we are analyzing the *availability* of crisis management capacities – we are not assessing their performance. The level of institutionalization suggests how widely known, trusted, and available these capacities are. It says little about their effectiveness.[21] A policy or practice may be very effective yet highly controversial, which hinders institutionalization. The opposite may happen as well: ineffective practices or organizations may persist much longer than expected because they are deeply institutionalized (Meyer and Zucker 1989).

To gauge the level of institutionalization within a policy sector, we must formulate indicators of institutionalization. We look for these indicators on three dimensions (cf. Stone Sweet, Fligstein, and Sandholtz 2001).

The first dimension relates to the *regulatory outputs* of a rule system. The EU formulates rules in various ways. Green papers and discussion briefs typically bear little influence, except in presenting new ideas and concepts. European directives and regulations, by contrast, are binding documents that prescribe formal action and require adjustment by member states. Of course, there is no direct correlation between binding law and behavior: powerful ideas presented in non-binding legislation can have an equal impact on behavior. The adoption of formal rules does suggest a higher level of institutionalization.

The second dimension pertains to *network configurations* within a sector. Some interaction patterns are loosely knit within a sector, representing the occasional gathering of national officials to share information and ideas.[22] Other interaction patterns are more tightly knit. Closely linked officials rely on those networks not just for information sharing but also for policy formulation.[23] The networks may acquire a "taken for granted" status, representing a level of trust and mutual dependence that marks the emergence of "security communities" (Adler and Barnett 1998). As networks move from loose associations to tight communities, the network takes on institutional characteristics.

The third dimension is directly related to the *level of legitimacy* a rule system enjoys and the level of authority it subsequently commands (Suchman 1995; Checkel 2001). A rule system or policy system gains legitimacy when its "products" – formal and informal prescriptions and prohibitions – are accepted without much debate by and within the member states. Council declarations, media commentaries, European Parliamentary debates, and opinion polls provide indicators – albeit rather "soft" ones – of the level of legitimacy that a European policy sector can boast.

To establish how and why the three crisis domains became institutionalized to the degree that they did, we have mostly relied on document analysis. We studied massive amounts of official documents, we accessed internal documents, and we made heavy use of the academic literature that analyzes various dimensions of EU governance. In addition, we interviewed scores of officials in EU institutions.[24] We provide detailed institutional sketches of emerging domains rather than fine-grained surveys of easily measured variables.

In our analysis, we have taken into account the major legal constraints, dispersed authority, and fragmented bureaucratic organizations that characterize the EU. These characteristics set the EU apart from a traditional state or international organization. The EU is a strange beast – neither a state nor an organization – which may limit the explanatory power of our theoretical perspective of choice.

A final word of caution: our assessment has very limited predictive value. As we write this, the financial crisis is still raging and threatening the institutional cement that has kept Europe together for decades. This and other future crises will test just how deeply the EU's crisis domains have been institutionalized. They will also affect the institutional trajectories of these domains.

Outline of the book

The message of this book is quite simple. The EU's crisis management capacities have progressively accumulated to a substantial and perhaps impressive degree – a fact that is somewhat surprising considering that crisis management remains a core responsibility of the nation state. Yet, the EU clearly has a way to go. Member states find it hard to agree on EU crisis management capacity "with teeth" as crisis management touches on a core national responsibility.

This is not a defeatist message, as found in other works (Etzioni 2007; Taylor 2007; Flood 2009). The EU is designed to enable close cooperation between member states and to coordinate their contributions. The EU is, in fact, very good at this. Its strengths, we argue, make it uniquely qualified to play a leading role in the joint response to crises and disasters that require a joint response. At the same time, the EU will have to "add value" to existing capacities and practices at the national level (Schmitter 2000; Anderson 2002; Joerges, Sand, and Teubner 2004; Dryzek 2006). It will have to offer complementary capacity that is useful but does not encroach on national sensibilities.

To substantiate our argument, we spend the following three chapters (2–4) describing how and to what extent the three crisis domains have become institutionalized. In each chapter, we answer the following questions:

(1) What are the policies that the EU has developed over time? Why were these policies developed?
(2) What resources are available? What organizations, people, funds, and instruments can be brought to bear by the EU?

(3) What leadership configuration is in place for each crisis type? Who makes the decisions, in which forums, and through what mechanisms?

These chapters take the reader on an intriguing journey across the EU's policy spectrum, exploring a world of legislative instruments, decision-making procedures, framework policy programs, new agencies, and generic administrative functions. After presenting our empirical findings, we offer an assessment of these capacities in Chapter 5. We will discuss how far the EU has come in time, but we will also discuss to what degree the EU has developed capacity that works. Finally, we offer some suggestions that may help to further improve the EU's crisis management capacities.

Notes

1 The official title is High Representative of the Union for Foreign Affairs and Security Policy.
2 Other agencies involved were the European Centre for Disease Prevention and Control (ECDC) and the European Food Safety Agency (EFSA).
3 Jörg Monar speaks of "systemic constraints on current EU crisis management action possibilities which could only be removed by a fundamental political and constitutional system change which at the very least looks rather unlikely at this stage" (personal communication, April 2010).
4 Subsidiarity is a governing concept, which holds that higher levels of administration should only assume tasks and competences that cannot be fulfilled at a lower level. Originating in the administrative philosophy governing the Catholic church, it has been a dominant concept in the history of the EU. The concept was formally included in the Amsterdam Treaty (1997).
5 See the 2010 report by the European Environment Agency for an overview of natural hazards and technological accidents in Europe.
6 Weberian bureaucracies are designed to deal with standard situations in a routinized and deliberate way (Wilson 1989). Uncertainty and urgency are the enemies of bureaucratic decision making. We use the term *bureaucracy* here not in a pejorative sense but as a reference to a particular form of organization common to the public sector.
7 On the reflexive nature of modern technologies, see Tenner (1995) and Rochlin (1999).
8 This point is nicely made by the EU's Reflection Group on the Future of the EU 2030 (2010).

9 The Hurricane Katrina disaster in 2005 provides a most telling example (Brinkley 2006; Cooper and Block 2006).
10 This definition builds on Rosenthal, Charles, and 't Hart (1989) and has become increasingly dominant in discussions of human, societal, and environmental security (Beck 1992; Buzan, Waever, and De Wilde 1998). See Rosenthal (1998), Stern and Sundelius (2002), and Boin (2005) for detailed discussions of the analytical advantages of using a generic crisis concept.
11 To be sure, there are crises that fit between these categories. It is, for instance, possible that Type I and Type II crises can escalate and snowball into Type III crises. In addition, there are many bilateral forms of crisis and disaster management, which aim to facilitate regional cooperation in response to emergencies in border regions (for example, cooperation between fire services in the border regions of Holland and Germany).
12 It is increasingly used to mobilize EU resources in response to crises and disasters occurring *outside* the Union.
13 For a detailed discussion of the complexities of disaster logistics, consult Boin, Kelle, and Whybark (2010).
14 This definition builds on the traditional notion of crisis with its core conceptual elements of threat, urgency, and uncertainty (Rosenthal, Charles, and 't Hart 1989; Rosenthal, Boin, and Comfort 2001). What sets it apart from the more traditional definition is its emphasis on the tightly woven web of critical infrastructures that characterizes modern society. For an elaborate discussion, see Boin and Rhinard (2008) and Ansell, Boin, and Keller (2010).
15 See the *International Herald Tribune*, "U.N. Panel says weather disasters are new norm," March 29, 2012, p. 1.
16 At the most general level, institutions can be defined as rule systems that structure social interaction by prescription and prohibition (Huntington 1968; North 1990; Goodin 1996). We are here particularly interested in organizational structures, rules, and practices.
17 In the EU, treaty-changing, "history-making" decisions often spark EU cooperation in new policy fields (Peterson 1995; Peterson and Bomberg 1999). Subsequently, the EU institutions engage in a process of fleshing out policy ideas and taking initiatives within the framework of the treaties. Lindberg (1965: 58), one of the doyens of integration theory, noted that structural, constitutional decisions in the EU tend to be "leaps in the dark" with unforeseen consequences.
18 This phase is sometimes referred to as the "institutional consolidation" period in between history-making decisions in the EU (Niemann 2006), during which "policy-shaping" ideas are floated and details fleshed out (Rhinard 2010).

19 EU institutional "experiments" rarely die, since termination typically requires the blessing of all member states (just one can prevent such a premature ending) and the acceptance of strong bureaucratic actors such as the Commission (which has an incentive to expand its competences).
20 We define a "policy space" as an institutional field of actors, rules, and practices associated with governmental efforts to address a particular category of social issues and problems (cf. Bourdieu 1990; Alink, Boin, and 't Hart 2001; Ekengren 2002). In the United States, this particular space is labeled as "homeland security." While the use of this term has increased in European policy circles, it is slightly controversial. We therefore avoid the term.
21 For studies on the effectiveness of EU external policy, see Ginsberg (2001) and Niemann and Bretherton (2013).
22 Such patterns reflect the type of loose "issue networks" identified in the public administration literature (Rhodes 1990).
23 This resembles the "policy community" category of policymaking networks (Christiansen 2006).
24 One of the authors worked part-time in the Brussels-based European Policy Centre (a think tank) for the past five years, which provided the team with access to a dense network of EU policymakers.

2 *Assisting overwhelmed states*

The evolving use of the Civil Protection Mechanism

The Union and its Member States shall act jointly in a spirit of solidarity if a Member State is the object of a terrorist attack or the victim of a natural or man-made disaster.

(Article 222 of the Lisbon Treaty)

Introduction: the gradual ascendance of the Civil Protection Mechanism

The earliest efforts to create joint crisis management capacity in the EU arguably date back to the 1970s. A set of now mostly forgotten disasters – who remembers Seveso and *Amoco Cadiz*? – prompted Commission officials to build what today is known as the Civil Protection Mechanism. This oddly titled organizational capacity helps member states deliver on the promise, captured in the Solidarity Clause of the Lisbon Treaty (cited above), of swift cooperation among national civil protection services in responding to natural or human-made disasters.

Since the first cautious steps were taken in the late 1970s, civil protection has emerged as a European policy domain – a "new European political space" (Stone Sweet, Fligstein, and Sandholtz 2001).[1] The EU has in place a set of procedures and instruments that allow member states to coordinate and pool their resources in the face of disaster. Today, civil protection is explicitly mentioned in the Lisbon Treaty, it is a recognized EU policy area, and its products, as we will see, are increasingly "consumed" across the world.

The emergence and rise of civil protection as an EU policy field is anything but self-evident. Civil protection – the activities aimed at protecting the population from the consequences of disaster – falls squarely within the traditional task domain of the nation state. No mention was made of it in the Treaty of Rome. To invest the EU with authority to deal with civil protection therefore touches on the sacrosanct principle of subsidiarity. The EU's founding fathers would probably be surprised to see how its member states have created a role for the EU in the area of civil protection.

One could, of course, argue that the EU has always been about protection (Boin, Ekengren, and Rhinard 2006). The very purpose of creating a European Community was to protect the member states, first, from warring with each other. In later years, the EU focused on protecting member states against market disturbances. But in the EU's early years protection was rarely, if ever, interpreted in the more tangible, perhaps physical, sense.

As we will discover in this chapter, the history of EU civil protection is best described in terms of "mission creep." After a slow start in the 1990s, member states strengthened and broadened the remit of this policy area after the dramatic events of September 11, 2001. The use of the Mechanism also changed: member states rarely call on it, but the Civil Protection Mechanism is increasingly being used *outside* the Union.

We start this chapter with a brief policy history. We then discuss several "signature events" that help us understand how the EU civil protection sector operates. We also sketch empirical patterns of the Civil Protection Mechanism's use. We analyze the institutionalization process of the Civil Protection Mechanism and conclude this chapter with an assessment of the middle-range future of this policy sector.

The Civil Protection Mechanism: a brief policy history

Two environmental disasters laid the foundation for what would become the EU's Civil Protection Mechanism. In 1976, a dioxin cloud escaped from a chemical plant in Seveso (Italy) and forced the evacuation of 600 homes. Nobody died (2000 inhabitants of the area were treated for dioxin poisoning), but the member states agreed afterwards that joint action was necessary to prevent similar and potentially more disastrous accidents (Wettig and Porter 1998).

The EU created an elaborate set of procedures (Seveso Directives I and II) that still govern the safety of chemical plants in member states.[2] The Seveso guidelines have helped member states to establish a minimal level of safety in their chemical industries (member states can point to a common source of regulatory pressure). The EU's Major Accidents Hazards Bureau (part of Joint Research Centre) offers scientific and technical support in the Commission's supervision of member state activities in this domain.

On March 16, 1978, the oil tanker *Amoco Cadiz* broke in two during a spring storm off the coast of Brittany, France. Polluting 320 km of French coastline, the oil spill resulted in a large loss of marine life and caused a public uproar. In reaction, the European Council decided that member states "must have very prompt access to information on the human and material resources which can be deployed for the control of such pollution."[3] The *Amoco Cadiz* disaster eventually (in 2002) gave rise to the European Maritime Safety Agency. This agency formulates guidelines for crisis response in the event of an oil spill.

The *Amoco Cadiz* disaster also gave rise to the more general idea that the EU might play a role in helping member states requiring expert assistance and critical resources in the management of a disaster. Subsequent efforts to create a civil protection mechanism would be modeled on this initial effort to create a marine pollution framework (Wendling 2010).

The Commission's reaction to the Seveso and *Amoco Cadiz* disasters was handled by the DG Environment, which over time would become a key player in the EU's civil protection sector (until the Lisbon Treaty moved the Mechanism to DG ECHO). The Commissioners holding the environmental portfolio at the time, Carlo Scarascia-Mugnozza (1973–7) and Lorenzo Natali (1977–81), were both Italians and thus intimately familiar with the devastation of natural disasters: Italy arguably has suffered the most from disaster in Europe. The Italian Commissioners led a drive to give the Commission more responsibility in pre-identifying resources in the EU and providing a "one phone call" source for swift deployment of these resources.

The underlying idea was that large-scale natural disasters that overwhelm a member state require collaboration among assisting countries. This collaboration, in turn, was thought to foster solidarity: it would "help establish a People's Europe" – "fostering in Europeans the sense of belonging to the same community."[4]

These early efforts did not build on the oldest source of EU capacity for disaster management. The Euratom Treaty (1958), one of the founding documents of the EU, provides the Commission with a unique legal instrument that has no parallel in the whole of European Community Law (Liberatore 1999: 200):

> In cases of urgency, the Commission shall issue a directive requiring the Member State concerned to take, within a period laid down by the Commission, all necessary measures to prevent infringement of the basic standards and to ensure compliance with regulations. (Art. 38 Euratom Treaty)[5]

The Mechanism started in a more cautious manner. In 1987, the Council called for "practical steps" and invited the Commission to "introduce a Guide to Civil Protection [which] will include a list of liaison officers from the Member States . . . so that the information collected will help to produce a clearer picture of the assistance available in each Member State in the event of a disaster."[6] Other "practical" (if modest) steps included the exchange of experts, the creation of a databank of existing systems and resources in member states, and the initiation of training programs. The next year, a "single additional emergency telephone number" was proposed.[7]

A 1991 Council Resolution contained the basic description of the Civil Protection Mechanism as it would look for the next two decades. It advised that member states "shall, if requested by another Member State, furnish all such assistance as they deem possible and available in the event of disaster in the territory of that other Member State" (see Åhman and Nilsson 2009: 85).[8] The resolution specified that assistance would take the form of "early dispatch of aid teams, supplied with equipment and aid material." It determined that "direction of the aid operations shall be the responsibility of the Member State requesting aid," which rules out any sort of managerial task for the Commission. It also explains how these "aid teams" should be received by the stricken country ("Member States shall endeavor to reduce to a minimum border checks and formalities for aid teams") and declares that "the costs of assistance . . . shall be borne by the requesting Member State."[9]

The role of the Commission remained purely facilitative. According to the resolution, the aid-offering and aid-receiving member states "shall forward to the Commission a report," which the Commission will pass on to other member states. In addition, "the Commission shall call meetings of the network of national liaison officers for the purpose of examining the technical and operational aspects of the organization of the cooperation provided for in this resolution."

The first real legal EU competence was not established until 1997. It took the form of an "action program," which directed the Commission to enhance its efforts aimed at the pooling of member state expertise, facilitating mutual assistance, and offering training programs.[10] By the end of the decade, the Commission was slowly building capacity to help carry out the intentions stated in the various Council declarations – but it still had very little to show in terms of actual policy successes.

The 9/11 attacks and the birth of the Civil Protection Mechanism

The events of September 11, 2001, created a sense of urgency in Europe: it was widely felt that the EU required a clear, coordinated disaster response strategy, which would ensure mutual assistance in times of disaster (which was now understood to include acts of terrorism as well). In October 2001, the EU officially established the Community Mechanism for Civil Protection to facilitate cooperation in civil protection assistance interventions.[11]

The idea of the Mechanism was, of course, not new and built directly on the activities described above. Momentum for an overhaul of the rather modest civil protection capacities in place had been building following coordination problems that plagued civil protection intervention teams operating in Turkey after the earthquakes in 1999. The events of 9/11 made such an overhaul look urgently necessary if not inevitable.

The Mechanism combined old ideas with a new set of tools, including a Monitoring and Information Centre (MIC), a Common Emergency and Information System (CECIS), Civil Protection Intervention Teams (based in the member states), and a database of available resources.

In one way, the Mechanism merely offers a "reinforced Community civil protection structure."[12] Viewed from an organizational perspective, however, it is clear that the 2001 Council decision enlarged the role of the Commission. Several new elements were introduced that would prove critical in subsequent years:

(1) The Mechanism offers protection in the event of emergencies "occurring both inside and outside the European Union" (which expands the territorial remit of EU civil protection).
(2) With respect to civil protection assistance outside the EU, the Mechanism could "also be a tool for facilitating and supporting

crisis management referred to in Title V of the Treaty on European Union" (which creates, or at least assumes, organizational links with other policy domains within the Union).

(3) The object of protection is more broadly defined, in terms of "people, but also of the environment and property, including cultural heritage, thereby reducing loss of human life, injury, material damage, economic and environmental damage, and making achievement of the objectives of social cohesion and solidarity more tangible" (which potentially enlarges the types of crises and disasters that may trigger the Mechanism).

(4) The Mechanism can be triggered in response to "a major emergency within the Community, or imminent threat thereof, which causes, or is capable of causing, transboundary effects" (which not only broadens the types of adversity the Mechanism may address but also implies an expanded organizational remit).

These new elements did not address a dilemma of almost existential dimensions: should the EU endow its Civil Protection Mechanism with operational capacity (which would touch upon the subsidiarity principle) or should it seek to use capacity already existing elsewhere within the EU (see Chapter 3) or even outside the EU (NGOs, NATO, UN)?

The Council continued its cautious trajectory. In the aftermath of the 2004 tsunami disaster, which exposed the limits of the Mechanism (see p. 36 ff.), the Council chose to strengthen the Mechanism by introducing so-called modules. Instead of each member state offering assistance on an ad hoc basis, a predetermined set of civil protection assistance modules, allowing for rapid deployment in case of a disaster, would be built. These modules are autonomous, self-sufficient units located in member states that can be moved to a disaster site on short notice.[13] The modules remain under the direction of member states.

In addition, the Council called for improved coordination between the Civil Protection Mechanism and the CSDP's (Common Security and Defence Policy) civil–military missions (for instance, appointing a Commission liaison to the Council's civil–military cell – see Chapter 3). But it postponed the possibility of fundamental reform, acknowledging that "in the longer term more ambitious and structural reforms of the Mechanism can be envisaged."[14]

A clarion call for further strengthening of the Mechanism came with the Barnier report, *For a European Civil Protection Force: Europe*

Aid (2006), which proposed the establishment of a European Civil Protection Force (Europe Aid). While the report was initially ignored, at least politically, the idea of an autonomous EU disaster management force gained popularity over time, especially in the southern member states.[15] After the forest fires in southern Europe in the summer of 2007, Greece called for a "genuine" operations center at the EU level, simplifications for EU co-financing of transport, and the creation of a European pool of forces coordinated by the MIC.[16]

These calls contained the seeds for change, but the immediate reaction was still cautious. A recast (revision) of the Community Mechanism was adopted by the Council on November 8, 2007.[17] The recast sets the parameters for the development of a European rapid response capability based on civil protection modules of the member states.[18]

The Council did establish a new legal basis for granting Community financial support to civil protection actions and measures within the EU: the so-called Instrument for Stability provides the legal basis for financial support to civil protection activities outside the EU.[19]

The EU also has in place financial capacity to help member states "bounce back" during the long aftermath of large-scale disasters.[20] The EU Solidarity Fund was created for the provision of financial assistance to help people, regions, and countries hit by major disasters to return to normal living conditions. The fund can be mobilized up to a maximum amount of €1 billion annually.[21] It is to be used for restoration of critical infrastructures such as energy, water and waste water, telecommunications, transport, health, and education. It can also be used for funding of rescue services, securing of preventive infrastructures and measures for immediate protection of cultural heritage, and the cleaning up of disaster-stricken areas. The Commission's DG Regional Policy is responsible for managing the fund. It has been used for 33 disasters covering a range of different catastrophic events such as floods, forest fires, earthquakes, storms, and drought. Twenty different European countries have been supported so far for an amount of more than €2.1 billion.[22]

The Solidarity Clause

The most legally prominent development has been the introduction of the Solidarity Clause, enshrined as Article 222 in the Treaty on the Functioning of the European Union (TFEU). This one-page provision creates one of the most explicit demands upon EU members to act jointly and to assist one another in the face of disasters, emergencies, and crises on the European continent. Yet the precise meaning of this demand, and its implications for EU institutions and member states, are still being assessed (Myrdal and Rhinard 2010; Fuchs-Drapier 2011).[23]

The concept of "solidarity" is one of the more symbolically powerful elements in the narrative accompanying European integration and has been invoked in recent years in the aftermath of ash clouds, pandemics, explosions, and cross-border infrastructure failures.

The Solidarity Clause contains three obligations. The first enjoins both EU institutions and member state governments to "act jointly . . . if a Member State is the object of a terrorist attack or the victim of a natural or man-made disaster." By including the EU institutions with national governments in this phrasing, the Clause suggests a supranational (rather than a purely intergovernmental) response, while the inclusion of broadly defined disasters indicates a fairly wide perspective on threats. The second obligation states that the "Union shall mobilize all the instruments at its disposal, including the military resources available by the Member States" to both prevent and respond to terrorist attacks and to assist one another in cases of terrorism or natural disaster. The third obligation demands that EU institutions and member state governments organize their responses via the Council of Ministers, thus suggesting a coordinated effort from the EU level when attacks happen or disasters strike (Art. 222, TFEU).

All this leaves ample room for interpretation. Solidarity means different things to different people and to different governments. For some, solidarity is measured by how much support flows to a country in need. For others, solidarity means everyone doing their "homework" to avoid the need for assistance in the first place. Still others believe that solidarity in the face of today's risks and threats is best pursued outside of EU frameworks. In the past, as long as "solidarity" remained a rhetorical concept, such differences could be tolerated. Now that solidarity has been established as a legal concept in the form of the Solidarity Clause, such differences need to be reconciled. Moreover, instruments to ensure a true "Union" response to crises will likely require new structures and procedures for inter-institutional and cross-sectoral cooperation (Myrdal and Rhinard 2010). The Treaty gave rise to a new Council Committee on Internal Security, COSI, which is being tasked with coordinating the implementation of the Solidarity Clause (see Chapter 4).

Expansion after Lisbon

The Lisbon Treaty boosted the reach and capacity of the Mechanism. This shift was set in motion in 2008, when the Commission announced a set of initiatives that built directly on the Barnier report mentioned above. These initiatives aimed to

(i) build up MIC to an "operational center" (which would require a "qualitative shift" from information sharing to operational engagement and coordination);
(ii) work toward a European civil protection force (developing reserve resources and standby modules);
(iii) develop logistical capacity (including stockpiling) for humanitarian disasters.

These initiatives have received a significant boost in the post-Lisbon era.[24] A newly initiated European Voluntary Humanitarian Aid Corps (fully operational in 2014) will provide operational capacity. Importantly, the MIC will be allowed to pre-position capacities made available by the member states if a disaster looms. In addition, the Commission will finance the development of the necessary transportation capacities. In short, one can see the outlines of an EU disaster force with operational capacities.

To support this budding force, the MIC has been moved from DG Environment to the DG Humanitarian Aid and Civil Protection (ECHO) and merged with ECHO's crisis room. The MIC now serves member states, other overwhelmed countries, and international organizations (the United Nations in particular). Moreover, the Commission has enhanced its role vis-à-vis the member states. Not only will it manage a process of certification and registration of capacities that member states make available, it will also have a stronger coordinative role of joint response operations (operations outside the Union will be coordinated by UNOCHA, the United Nations Office for the Coordination of Humanitarian Affairs).

The Mechanism's scope and reach is being expanded by including risk management and disaster prevention in its remit. Observing "a need for, and significant added value in, European action in the area of disaster prevention," the EU had already begun to explore what it can do to support and enhance member state actions to prevent disasters from happening.[25] For instance, the Commission makes use of early warning systems such as the European Flood Alert System and the European Forest Fires Information System (EFFIS),[26] both developed by the EU's Joint Research Center (JRC).[27]

The activities in the civil protection area build on a strong belief that national systems must converge and be improved to facilitate a joint response to crisis and disaster. To that effect the Commission has introduced guidelines on risk assessment, which provide a basis for national disaster risk management plans.[28] Member states have to send these national disaster risk management plans to the Commission.

Some of the original aims that had received less attention over the years have been revived. In the original documents, for instance, a pronounced aim was to "contribute to public information to help European citizens prepare for disasters," but this never materialized as a major task (or a marked success). These documents also promise to help member states "prevent and prepare" for disasters. Another pronounced aim was to initiate public information and education campaigns with the objective of increasing the degree of self-protection. Finally, a stated aim was that member states learn from each other. The post-Lisbon changes reiterate these aims and provide the Commission with additional (if modest) means to accomplish them.

The Mechanism: how it works

At the heart of the Mechanism is the EU Civil Protection Monitoring and Information Centre (MIC). Originally located in the Civil Protection Unit of the Commission's DG Environment, it moved to DG ECHO in 2012 following a Commission reorganization.[29]

The MIC operates on a 24/7 basis. The core task of the MIC is to facilitate cooperation between member states in response to a disaster. The Mechanism was originally intended to deal with natural disasters (earthquakes, floods), but the definition has widened over time to become virtually all-encompassing:

> There is no black-and-white definition of disasters and I think that is a very deliberate choice. In the end, it is up to each Member State to see,

and to determine, when it wants to make an appeal for assistance and for which particular situation. The key factor in all of this is that, at some point in a crisis, national coping capacities and national response capacities may become overwhelmed, and the point where that is reached will be very different from Member State to Member State and it will certainly be different inside the EU from outside the EU, so a disaster which may be perfectly manageable for one country may become unmanageable for another country. (Testimony of Hans Das, Head of Civil Protection, quoted in House of Lords, 2009a: Q7)[30]

Any country – member state or not – can request assistance from the EU when it is overwhelmed by a disaster. All assistance is, in principle, voluntary in nature.[31] Voluntary assistance has obvious limitations. When, for instance, Portugal asked for assistance to deal with its forest fires, 21 countries expressed a clear political will to help. Yet, Portugal received assistance from only Italy (two Canadairs) and Germany (three helicopters) because most available capacity was already being used to fight fires elsewhere in the Mediterranean region.[32]

The MIC uses a web-based system (CECIS) to log requests for assistance and distribute these requests to member states.[33] When a member state requests assistance through the Mechanism, the MIC has, in effect, little to do (this will likely change as the MIC gains more operational capacity and coordination powers). The requesting party contacts other member states through the system and the teams that come in to help are entirely managed by the member states. The MIC trains member state representatives how to use the system.

When non-members request assistance, the MIC has more tasks. Requests from outside the EU must first be vetted by the EU Presidency to see if it is a "clear-cut" disaster or a crisis situation (which might require the use of CSDP capacity). Foreign countries cannot access CECIS by themselves. The MIC "does so on behalf of the foreign country in the system." In addition, Mr. Das explains, the MIC has an oversight role:

When it is an emergency outside the EU, in particular in developing countries, but also in other third countries, we, as a rule, always send assessment and coordination teams to make sure that we are sending assistance which meets the needs on site and which is really in accordance with the needs identified by our experts, and we send coordination people to make sure that the European assistance is correctly received and that it is correctly distributed in full coordination with other relief actors.

The MIC plays two additional roles. First, it monitors ongoing disasters and new disasters. The MIC is connected to the JRC's early warning system and receives information from other international organizations (such as UNDAC) about unfolding disasters. Second, the MIC attempts to

"make sure that Member States have always up-to-date, reliable information on what is going on during emergencies.... We find it extremely important that we can inform our Member States very early on about the possibility of new disasters or new disasters happening which may develop into requests for civil protection assistance." (Testimony of Hans Das, House of Lords, 2009a: Q4)

With the post-Lisbon enhancement of the MIC's administrative capacity set to take effect in the upcoming years, its activities are likely to expand. In fact, the MIC may well begin to play a role resembling the one played by FEMA (the US Federal Emergency Management Agency): actively pre-positioning emergency supplies in preparation for an impending disaster and assuming a central position in the multi-actor response network.

A network of experts and resources

An explicit aim of the Mechanism is "to play a coordinating role between civil protection experts" from the member states to "ensure that the best personnel are dispatched to disaster areas as quickly as possible."[34] To this effect, a Permanent Network of National Correspondents (PNNC) has been created.[35] This operational network of civil protection experts serves as a forum for exchange of information and discussion of different initiatives in the field of civil protection. The network encompasses contact points in all EU member states (as well as the former Yugoslav Republic of Macedonia [FYROM], Iceland, Liechtenstein, and Norway).

To help prepare national civil protection experts to provide assistance within and outside Europe, the Commission has offered specialized training courses annually since 2004 (well over 1000 national practitioners have been trained). The courses are intended to foster "experience-sharing and networking between national experts... where experts can learn first-hand about similar responsibilities under different national systems."[36] The main focus is on operational procedures and familiarization with different national structures, concepts, and approaches.

The Mechanism uses two additional tools to ensure "complementarity and compatibility between intervention teams coming from different Mechanism countries" (EVA4CP 2009): simulation exercises and the exchange of national experts (469 exchanges had been realized by 2008). As of 2012, the Commission had funded 30 EU-wide simulation exercises.[37]

There are only limited financial means available for the Mechanism. For 2010, the EU's civil protection budget was €26 million and an additional €7.5 million was available to further create national modules.[38] The budget for the years 2014–20 envisages budgetary commitments for €276 million for use inside the Union and €237 million for use outside the Union (which is a slight increase).

Use of the Civil Protection Mechanism: empirical observations[39]

The EU frequently mentions successful uses of the Mechanism. Just hours after the 9/11 attacks, for instance, a team of 1000 emergency workers was reportedly ready to fly to the United States. Another oft-cited success story mentions the request by the Czechs – then not part of the EU – for water pumps in 2002, after the country suffered major floods. Its use has been reported in response to terrorism (Madrid 2004; Mumbai 2008), the forest fires in the EU's southern regions (2003, 2004, and 2007), faraway disasters (Asian tsunami; Hurricane Katrina; Haiti), and foreign hot spots (Lebanon 2006).

Within the 2002–11 period the Mechanism was activated 157 times.[40] Figure 2.1 illustrates the growing trend in the number of activations and reveals that a slight majority (60%) of the activations was by non-EU countries. The EU member states that suffered the most from disaster were not necessarily the most prolific users. For instance, Bulgaria lodged more requests for flood assistance than France or the UK (which suffered more flooding). The richer EU members appear to make less use of the Mechanism.

While it is true that use of the Mechanism has increased over the years, the use of the Civil Protection Mechanism may well be subject to some inflation in the official documentation. Even the smallest missions (consisting of two or three members) are listed, as are missions that were offered but not accepted by a foreign country. The MIC is publicized much more in EU documentation (not particularly

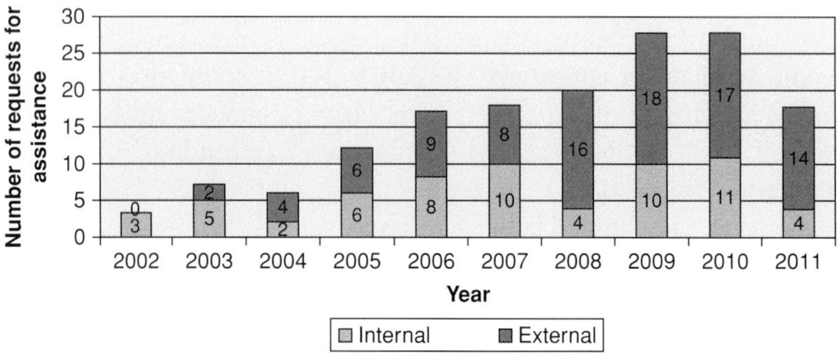

Figure 2.1 Requests for assistance inside the EU and in third countries

surprising) than in national documents. For instance, in a Swedish Foreign Ministry paper on the management of the tsunami crisis, the MIC is mentioned in only one paragraph. The EU's own reporting very distinctly mentions the use of the MIC.

So what exactly can the EU do to facilitate a joint response to disaster-struck countries? To get a more precise idea of the MIC's role, we briefly revisit four "signature crises" that helped shape the Mechanism and examine how it is being used today.

The floods in Central Europe (2002)[41]

In the summer of 2002, large parts of Central Europe were flooded. The Czech Republic was among the worst-hit countries: 40% of Czech territory was affected, 6% was inundated; in Prague, 50 000 residents had to be evacuated. The Czech Republic, which at the time of the floods was not yet a member state, requested assistance on August 14, 2002. DG Environment (MIC) activated the Community Mechanism accordingly.[42]

When the request for assistance was received, the crisis had already taken on major proportions. The Community Mechanism had only existed for six months at the time of the flooding; not many countries were yet aware of its existence. In fact, it was the Commission that contacted the Czech authorities and urged them to request assistance via the Community Mechanism. The MIC served as the central hub for information: the center was continuously updated with news on the flooding, over the telephone, and in written form (the web-based CESIS did not exist then). The MIC, in turn, informed EU member states and other Central European countries affected by the disaster.

The MIC linked Czech requests for help to offers of assistance from EU member states. The Czechs requested portable dryers, floating pumps, and electric submersible pumps. Material and personnel resources were provided by several EU member states (France, Belgium, Italy, and Greece). Since the use of the MIC is voluntary, countries did not always inform the MIC when assistance was offered on a bilateral basis. This, at times, caused confusion. The Commission offered financial support by reallocating resources from the EU structural funds (for Germany and Austria) and pre-accession funds for the candidate countries.

The sinking of the Prestige[43]

On November 13, 2002, the *Prestige*, a 26-year-old single-hull tanker, carrying 77 000 tons of heavy fuel oil, sprang a leak off the coast of Galicia, Spain. The tanker broke apart on November 19 and sank 270 km off the Spanish coast, spilling tons of oil into the sea. The spill caused major damage to the environment, polluting more than 200 km of Spanish and French coastline. It had a disastrous effect on fishing and shellfish farming along the Atlantic coast.

The Spanish authorities informed the Commission immediately and the Community Mechanism was enacted in response. Together with the JRC, and in coordination with the Spanish authorities, the MIC acquired satellite images of the area struck by the disaster. These images were then analyzed by the JRC so as to follow the drift of the oil spilled and to determine the damage.

The MIC mobilized resources from member states to help the cleanup operation. Spain instructed the Commission to first approach three individual member states (France, the Netherlands, and the United Kingdom) before launching the request to all member states. Both France and the Netherlands sent specialized vessels.

The Commission adopted special emergency measures and made resources available to help the Galician fishermen. Funds were made available through the "Financial Instrument for Fisheries Guidance."

To make this possible, DG Fisheries and Maritime Affairs drafted a proposal. Spurred by a sense of urgency and political agreement, the Council agreed within days.

Portuguese forest fires in 2003[44]

Portugal's summer of 2003 was one of record temperatures, well above 40 degrees Celsius, and strong winds – ideal conditions for rapidly spreading wildfires. By August 2003, 1700 forest fires had ravaged the Portuguese countryside. The worst fires in 23 years, they caused the deaths of 20 people. Portugal lost nearly 6% of its forest area. More than 3000 firefighters, aided by soldiers and officers of the National Republican Guard, battled the widespread fires with some 800 vehicles, water-carrying aircraft, and helicopters.

On the night of July 31, 2003, Portugal requested help from the European Union through its National Service for Fire and Civil Protection. Portugal issued a request for water-carrying aircraft and received an initial list of available resources from member states on August 1. A more complete list followed on August 2.

The member states provided two Canadairs (Italy), several water-carrying aircraft (France, Norway), direct aircraft assistance (Spain and Morocco), and three Puma helicopters plus crews (Germany). When fires simultaneously erupted in Spain and Italy, both countries had to recall their aircraft to fight forest fires on their own territory. In the aftermath of the forest fires, which caused damage estimated at €1.4 billion, the European Parliament approved emergency recovery aid of €48.5 million to Portugal from the EU Solidarity Fund.

The 2004 Asian tsunami[45]

On December 26, 2004, an earthquake measuring 9.0 on the Richter scale caused tsunamis that devastated South and Southeast Asia (Indonesia, Thailand, Sri Lanka, India, the Maldives, Bangladesh, and Myanmar) and East Africa (Tanzania, Somalia, and Kenya). The disaster is believed to have caused the death of between 228,000 and 310,000 people. Indonesia and the northern part of Sumatra, Aceh, suffered the highest death toll.

The EU had almost 2000 citizens either missing or confirmed dead. One of the countries most affected within the EU was Sweden, which had more than 30 000 tourists in the region and lost around 550 citizens in the disaster.

As this was a disaster outside EU borders, various EU entities were involved. In addition to the MIC and DG ECHO, DG External Relations played a role (its crisis room coordinated contacts with the EU Presidency and member states). Most of the ECHO efforts, as well as the vast majority of the MIC efforts, aimed at providing medical personnel and equipment, hospitals, sanitation, food, water, water purification tools, and temporary shelters.

The JRC was notified of the earthquake through its alert system, which had been triggered by a US Geological Survey bulletin, one hour and 20 minutes after the event had occurred. The JRC then immediately alerted the European Union's humanitarian aid department (ECHO), the Council's Joint Situation Centre (SitCen) crisis room, and the United Nations. DG ECHO established contact with the United Nations Office for the Coordination of Humanitarian Affairs (UNOCHA), which had taken on the role of main coordinator of humanitarian aid on site.

The MIC was alerted at 8.45 a.m. CET (Central European Time). The MIC contacted the EU Presidency, at that point held by the Dutch government. DG Environment's Civil Protection Unit started to communicate with the UNDAC team in Colombo, Sri Lanka, on December 26 regarding needs assessment and coordination of aid. At 10.00 CET, after an official appeal from Sri Lanka, the MIC sent a call for assistance to the member states. The first assessment teams were in place in Thailand and Sri Lanka on December 27.[46]

The list of initial assistance offered by member states in response to the Sri Lankan call for help included a support team of 60 people, sanitation equipment, medical aid, tents, a team of 17 doctors and staff, two Hercules aircrafts, and relief shipments. Teams of European firefighters were sent directly to the tsunami-hit region.

The receiving countries included Indonesia and the Maldives (in addition to Sri Lanka and Thailand). Stavros Dimas, then Commissioner with responsibility for civil protection, emphasized in a press release that this was "one of the most significant relief efforts in the history of the Community Civil Protection Mechanism."[47]

On January 5, the first teams coordinated by the Community Mechanism began to return to Europe. They were firefighters, and the reason for leaving the affected area was that the initial emergency phase of the crisis management was then considered to be over. From that date on, the focus would be on longer-term humanitarian aid.

Analysis: institutionalization in slow motion

At first sight, the level of institutionalization for the EU's Civil Protection sector may appear "low." It took a long time for the EU to develop a rather simple coordination mechanism (a computer system, a small crisis room, and an outsourced training program) that has been rarely used by its intended audience (overwhelmed member states). Apart from a few pumps and the occasional airplane, no member state has relied on the Mechanism for essential assistance in response to a super disaster (our Type I crisis). The Civil Protection Mechanism, we might say, is a solution looking for a problem.

The absence of overwhelming disasters within the EU may explain this limited use. But even when real disasters occurred in member states, the Mechanism was used only sparingly. It appears that the larger member states in particular are loath to invoke the Mechanism, which suggests that its legitimacy is not very high.

Another indicator of low institutionalization is the ongoing mission shift, from a simple instrument for coordinating mutual assistance to building operational capacities that are on standby and can be used around the world. This shift can be characterized along several analytical dimensions:

Geography: outside v. inside the Union. What are the boundaries of EU civil protection? Does it apply to member states only, to citizens of member states (wherever they travel), or to all areas in the world struck by disaster? Over time, the focus has shifted from an exclusive orientation on helping member states to providing assistance around the world.

Adversity type: specific disaster v. all hazards. Does EU civil protection pertain to specific types of adversity (such as natural disasters and terrorist acts) or to all types of hazards? The shift has been toward an all-hazards perspective.

Added value: coordination v. management. Does EU civil protection, in accordance with the subsidiarity principle, merely coordinate member state assistance to overburdened states, or does it aim to fulfill the mandate by offering operational capacity to stricken areas? The trend has shifted, albeit cautiously, toward a more proactive stance.

Activity type: disaster management v. humanitarian aid. What, exactly, is civil protection? Does civil protection pertain to all phases of disaster management (including prevention, mitigation, and aftermath) or does it primarily entail the provision of assistance to stricken populations (humanitarian aid)? Here, again, we observe an expanding notion of what the EU's tasks should be.

Crisis phase: only response v. full cycle. Does the EU only concentrate on the immediate response to a disaster, or does it include the prevention and aftermath phases of the crisis management cycle? In recent years, the shift has been toward embracing disaster prevention as a European project.

But this is not the complete story.

While the original remit of the Mechanism has been subject to expansion, there also has been convergence toward a comprehensive and widely agreed-on conception of what civil protection should be and for whom it is intended. This combination of rising ambition and policy convergence is reflected in, and supported by, the Lisbon Treaty and subsequent decisions of the European Parliament and the Council. Moreover, while it was sparingly used for its intended purpose (helping overwhelmed member states), the Mechanism is increasingly being used for the coordination of joint missions *outside* the EU:

> What started as a vehicle for assistance between Member States . . . has become a much broader mission for using exactly the same capabilities for exactly the same purposes – to save human life – but across a much broader piece because there is no distinction between home and abroad any more. (Testimony of Bruce Mann, House of Lords 2009a: Q48)

In terms of our institutionalization cycle, it appears that the Mechanism as originally envisioned did not get past the initial phase of institutionalization as there simply was no need for what the Mechanism purported to offer. The first institutionalization cycle did not come very far or run very deep. But, as if guided by an invisible hand, the Mechanism searched and found an alternative client base. As a result, the EU's Civil Protection Mechanism finds itself at the very beginning of a new institutionalization cycle, which has enjoyed a flying start.

The Mechanism appears to be of much better use in imposing coherence on the EU's effort to help faraway countries deal with massive disasters (the likes of which we have not seen in Europe). And the Mechanism is being strengthened. In forming his new Commission, Barroso shifted the MIC to the new Humanitarian Aid and Civil Protection portfolio headed by Commissioner Kristalina Georgieva. This will likely strengthen the EU's capacity to initiate a coordinated response in the face of international disasters. Let us now briefly recapture the Mechanism's history in terms of our institutionalization cycle.

A simple tool for an overwhelming problem: the roots of mission shift

A core problem – in terms of institutionalization – was the initial mismatch between a simple tool and its purported aim, which was highly ambitious but perhaps somewhat superfluous. National disasters are rarely so big as to require outside intervention. The Mechanism is thus only likely to be truly useful in times of *super disasters* – the likes of which we have not seen in decades (think of the 1953 floods in the Netherlands). Yet, the Mechanism as initially envisioned has little to offer in times of such disasters. When the critical infrastructures of a nation are paralyzed (think of the 2010 Haiti earthquake), a database of available resources will only provide the beginning of what will likely be a huge rescue operation.

It thus makes sense that the Mechanism was little used, which, in turn, minimized the opportunity to demonstrate its added value. When it was used, its impact was trivial ("coordinating" the request for a few water pumps or planes, which could easily have been done without the MIC).

Rather than killing the Mechanism, policymakers began to expand its goals. The second action program, which ran from 2000 to 2004, widened the activity repertoire to include risk evaluation, prevention, and mitigation; information to the public (creating a universal language for the traveling EU citizen, a common 112 number); and preparedness (improving disaster medicine, psychological aftercare, improving training conditions for first responders). The activities have little to do with coordinating assistance to an overwhelmed country.

Not only did the EU widen the object of protection from people to environment, property, and cultural heritage – it explicitly announced that the Mechanism could become a tool for CSDP use (see Chapter 3). This widening of the object of protection would have profound consequences for the character and use of the Mechanism.

European policymakers slowly became convinced that "European citizens expect the Union to protect their lives and assets inside the EU, and at the same time deliver effective disaster assistance in other parts of the world as an important expression of European solidarity."[48]

The Eurobarometer offers support for their conviction. There is widespread support for the idea of the EU offering civil protection assistance. A whopping 89% of a 2012 poll of EU citizens agrees that the EU needs a civil protection policy and that the EU should help any non-EU country worldwide hit by disasters.[49] Yet it is unclear why this aim would require "a comprehensive approach to the continuum of disaster risk assessment, forecast, prevention, preparedness and mitigation (pre- and post-disaster), bringing together the different policies, instruments and services available to the Community, and Member States working as a team."[50]

At the same time, member states hesitated to move the Mechanism beyond a mere phone book of national contacts and a database of national resources. When the Commission proposed to turn the MIC into a fully operational center, which would require a "qualitative shift" in its activities and processes, member states balked.[51] Mr. Das, the leading MIC official, quickly downplayed that ambition: "Very clearly, the development of the role played by the MIC in that respect does not involve substituting operational competences of the Member States; nor does it involve creating any operational command powers of the MIC over the experts in the teams made available by the Member States" (House of Lords 2009a: Q31).

Emerging problems

A first realization that the Civil Protection Mechanism was both potentially important and underdeveloped came in the wake of the 9/11 attacks. Policymakers across Europe understood that a major attack (a terrorist attack using smallpox was widely feared in the post-9/11 years) would require a coordinated response of member states. The Civil Protection Mechanism was not designed for that purpose.

The Asian tsunami (December 26, 2004) put the Mechanism to its first test. It was a major disaster in which many European citizens from different member states were directly affected, demanding a coordinated effort from EU countries.

Not surprisingly, the joint response did encounter its share of coordination problems – the tsunami, as the EU euphemistically put it, "brought home both the operational and political complexity of disaster and crisis preparedness and response." For one thing, the EU lacked logistical capacity to move resources to the region. The challenge, as the EU learned, is to "ensure an efficient, rapid and flexible response in the face of emergencies as well as to mobilize and develop these resources in a coherent manner."[52] The EU, in other words, discovered that a database of available resources in member states only goes so far. An effective, multinational response requires operational capacity. This the Civil Protection Mechanism did not provide for.

An effective joint response in a faraway arena required the participation and cooperation of DG ECHO. On paper, the difference is fairly clear. The aim of the EU's Civil Protection Mechanism is to provide assistance in the intervention phase. The Mechanism aims to send out EU experts within 12 hours of an event. DG ECHO, on the other hand, was mainly involved in the recovering phase and the rebuilding of infrastructure. Another difference pertains to the type of assistance. The Mechanism pertains to resources of member states, whereas ECHO mainly works with non-governmental organizations (such as UN Disaster Assessment and Coordination (UNDAC), the UN Refugee Agency (UNHCR), UNOCHA, and the Red Cross). Civil protection assistance is provided only in response to a request from the relevant state. Humanitarian aid through EU-ECHO provides "apolitical," non-discriminatory humanitarian emergency assistance regardless of any request or agreement from the country hit by disaster.

But the relation between the MIC and DG ECHO remained ambiguous (Åhman and Nilsson 2009: 90). This was due to the blurring of task domains: while the Civil Protection Mechanism was originally intended for internal use (assisting member states), it was increasingly used in response to international disasters – the traditional "domain" of DG ECHO.

One might expect cooperation between Commission entities, such as the MIC and DG ECHO, and Council institutions that planned faraway CSDP missions (see Chapter 3). But such cooperation has been marred by inter-institutional strife and a fundamental disconnect in perspectives.

A 2011 impact assessment concluded that the Civil Protection Mechanism suffered from a set of long-standing problems: reactive and ad hoc mechanics that limit the effectiveness, efficiency, and coherence of European disaster response; capacity gaps; limited transport solutions; heavy procedures; and limited preparedness.[53] This list is long enough to stop any institutionalization momentum in its tracks.

Adaptation: slow but effective

Over the years, the Mechanism has been adapted in two ways: there has been a gradual task expansion and the Mechanism has been made available for use outside the Union. This raised the ambitions (and possibly expectations) without addressing observed shortcomings.

When the Asian tsunami exposed the limited capacities of the Mechanism, the Council cautiously strengthened the Mechanism by introducing modules (discussed above). But the Council postponed the possibility of fundamental reform, acknowledging that "in the longer term more ambitious and structural reforms of the Mechanism can be envisaged."[54]

In December 2007, the member states, European Parliament, and Commission adopted the European Consensus on Humanitarian Aid, which is an explicit declaration of shared objectives and principles that underpin humanitarian aid. It underlines the need for strengthened coordination between Community and member states' responses to major disasters and in the domain of humanitarian aid policies.[55] It also offers a framework for improved delivery.

The Lisbon Treaty spurred further changes, but the EU's response to the Haiti disaster demonstrated once again that the Mechanism can only do so much when it comes to the quick mobilization of a massive response to a large-scale disaster.

The most recent changes (yet to be fully implemented at the time of writing) promise a real upgrade of operational capacities. The merger between the MIC and the DG ECHO crisis center into the European Emergency Response Centre (May 2013) should enhance the EU's capacity to coordinate a joint disaster response. Finally, a more structured relation with the EU's internal security sector (see Chapter 4) may enhance the EU's capacity to serve its original aim of assisting overwhelmed member states. It remains to be seen whether these proposed changes will make an impact in the long run.

Limited enthusiasm, low legitimacy

Who values the Mechanism? Few member states make use of it. The Madrid terrorist attack (2004) underlines this conclusion. Several member states offered immediate assistance through the MIC. Offers included field hospitals and beds for victims suffering from burn injuries. Spain, however, did not take up any offers of assistance. In the case of the London bombings, the authorities apparently did not even consider using the Mechanism, which they viewed as an instrument for use in response to natural disasters.[56] In both cases (Madrid and London), it was clear that the affected countries could deal with the consequences of the attacks and the only role that the MIC played was in channeling validated information from the authorities of the affected countries to all of the other member states.

The future of the Mechanism clearly lies with its use abroad, where it can help imposing some order on the chaotic and uncoordinated response of member states to foreign disasters. The use of the Mechanism outside of Europe is politically less sensitive. Few member states object to its use abroad (contrasted with the hesitation of some countries to use EU assistance domestically).[57]

Viewed from this perspective, the transfer of MIC to DG ECHO seems a smart decision. Not only does it eliminate a source of overlap, it also strengthens Commission capacity where the need is most dire. It does open an old source of tension: the relation with NATO. In its investigation of the EU's Civil Protection Mechanism, the British House of Lords observed a "most lamentable overlap of powers and activities between NATO and the EU" when it comes to mutual assistance in the case of disaster (House of Lords 2009a: Q63). Member states with a transatlantic leaning may well question the added value of the EU activities in this international domain.[58]

Conclusion: a tool looking for a goal

This chapter has described the history of the EU's Civil Protection Mechanism in terms of gradual evolution and "mission creep." With the initiation of the Civil Protection Mechanism, the EU intruded into a policy arena that has always been considered a core part of national sovereignty. It has done so slowly and with respect for subsidiarity concerns. While one can say that the Commission has made the most of its limited mandate, it has generated little enthusiasm for the Mechanism.

The successes it enjoyed depend mostly on its use abroad. The Mechanism has been used across the world in response to a wide variety of disasters (to "enhance coherence at the international level" has become an increasingly important aim).[59] Initially created to facilitate assistance to overwhelmed member states, the civil protection mechanism now most effectively serves EU disaster response efforts *outside* of the Union.

All this leads up to the question of whether there is a role for the EU before and during Type I disasters. The EU has the capacity to play a role, albeit a minimal one at this point. And there are additional capacities that could be "turned inward." But it remains unclear when a member state would really need the EU's assistance – and, when that happens, what the EU could offer to make a real difference.

As we will see in Chapter 4, a decade of policy expansion has culminated in a complete "rethink" of internal security in the Union. The Lisbon Treaty's Solidarity Clause suggests a new security system is emerging within the EU – a domestic-like system of crisis and disaster management. Civil protection, logically speaking, would play a role in this new system. Recent adaptations suggest this may happen, but the Mechanism is still – or again – at a very early stage of institutionalization.

It appears that the Mechanism has become a tool for the EU's foreign ambitions. How that came about we will discuss in the next chapter. But its status as a tool, rather than an institution, makes it vulnerable to shifting preferences and embarrassing failure. Intriguingly, the Union has been much more successful in managing the institutionalization process in what is widely considered a much tougher domain. It is to this domain of external crisis management that we turn next.

Notes

1 Tellingly, Stone Sweet, Fligstein, and Sandholtz (2001) did not recognize internal safety or civil protection as a new policy space or EU core

competence at the time of their writing. As we will see in this chapter, this policy space has expanded rapidly since they published their book.

2 See Council Directive of June 24, 1982, on the Major-Accident Hazards of Certain Industrial Activities (82/501/EEC). The Seveso directives do not apply to nuclear power plants.

3 Council Resolution of June 26, 1978 (setting up an action program of the European Communities on the control and reduction of pollution caused by hydrocarbons discharged at sea). This program would eventually result in the Community framework of cooperation in the field of marine pollution (2000).

4 European Commission (2000). "Council Resolutions in the field of Civil Protection: Developments since their adoption." Commission Staff Working Document. SEC (2000) 136. Brussels: January 24, 2000.

5 This Article 38 has never been applied, not even in the case of the Chernobyl disaster.

6 "Resolution of the Council and the representatives of the Governments of the member states, meeting with the Council of 25 June 1987 on the introduction of Community Cooperation on Civil Protection." Official *Journal of the European Union C 176.*

7 Council Resolution February 13, 1989, on the new developments in Community cooperation on civil protection. (89/C 44/03).

8 This may be viewed as an early predecessor of the Solidarity Clause (see page 27ff).

9 "Resolution of the Council, 8 July 1991," 91/C 198/01. See also "Resolution of the Council 23 November 1990," 90/C 315/02.

10 The second action program ran from 2000 to 2004.

11 Council Decision of October 23, 2001, establishing a Community mechanism to facilitate reinforced cooperation in civil protection assistance interventions. 2001/792/EC, *Official Journal of the European Union*, L 297, November 15, 2001, pp. 7–11.

12 Council Decision of October 23, 2001, establishing a Community mechanism to facilitate reinforced cooperation in civil protection assistance interventions. 2001/792/EC, *Official Journal of the European Union*, L 297, November 15, 2001, pp. 7–11.

13 The Mechanism countries are required to register modules and technical assistance support teams (TASTs) with the Commission. A module consists of "intervention teams" with specific skills that can be quickly mobilized in the case of a major crisis that demands the use of such a module. The implementing rules laid out in a Commission Decision in 2007 sets out 13 different module types which range from "Urban Search and Rescue" to "Advanced Medical Posts with Surgery." The

TASTs are designed to assist assessment and coordination teams, provide an on-site operations coordination centre, or be combined with another module. Commission Decision of December 20, 2007, amending Decision 2004/277/EC, Euratom as regards rules for the implementation of Council Decision 2007/779/EC, Euratom establishing a Community Civil Protection Mechanism (Annex III).

14 Communication from the Commission to the Council, the European Parliament, the European Economic and Social Committee, and the Committee of the Regions: "Reinforcing EU Disaster and Crisis Response in Third Countries." COM (2005)153 final. Brussels: April 20, 2005.

15 Barnier's remit was to report on "the EU's response to major cross-border emergencies" – "especially to major emergencies outside the EU." Barnier has since been appointed Commissioner in the second Barroso Commission (2010–14).

16 Observation based on interviews (held in 2007) with members of the Permanent Representations to the EU.

17 Council of the European Union (2007). Council Decision of November 8, 2007 establishing a Community Civil Protection Mechanism (recast) (2007/779/EC, Euratom) *Official Journal of the European Union* L 314/9. 1.12.2007 CHAPTER III Response Article 6.

18 It also renames the Community Mechanism "A Community Civil Protection Mechanism."

19 The Instrument for Stability provides for "exceptional assistance measures" to enable timely responses to disasters in addition to political crises outside the Union (see Chapter 3). See Proposal for a Council Regulation establishing a Rapid Response and Preparedness Instrument for major emergencies. COM (2005) 113final. 2005–0052 (CNS). Brussels: April 6, 2005.

20 The EU speaks of a "major disaster" when the damage is estimated either at or above €3 billion or more than 0.6% of the gross national income of the state concerned. Under exceptional circumstances – such as the floods in Central Europe in the summer 2002 – subregions can also request assistance from the fund.

21 Council Regulation (EC) No 2012/2002, of November 11, 2002 establishing the European Union Solidarity Fund. *Official Journal of the European Union* L 311.

22 http://ec.europa.eu/regional_policy/thefunds/solidarity/index_en.cfm (accessed February 3, 2013).

23 In April 2011, the European Parliament asked the Council to apply the Solidarity Clause to help Italy manage the situation in Lampedusa,

which had to cope with an exceptional number of illegal immigrants at that time.

24 "Proposal for a Decision of the European Parliament and of the Council on a Union Civil Protection Mechanism." COM (2011) 934 final. Brussels: December 20, 2011.

25 "Commission Staff Working Document" 7075/09 (quote taken from page 6) and "Commission Document SEC (2009) 202." This shift toward disaster prevention is further codified by inclusion in the Stockholm Programme.

26 The system consists of two modules: the European Forest Fire Risk Forecasting System, with fire risk forecasts for one and three days; and the European Forest Fire Damage Assessment System, with evaluation and damage surveys of fires covering 50 hectares or more.

27 The EU's Joint Research Centre (JRC) supports several ongoing projects pertaining to early warning. One of JRC's activities is "disasters and response." As such, JRC supports activities to develop better early warning capacity.

28 "Risk Assessment and Mapping Guidelines for Disaster Management." Commission Staff Working Paper, SEC (2010) 1626 final. Brussels: December 12, 2010.

29 The MIC's crisis room will be merged with the crisis room of DG ECHO. This merger is not expected to change the activities and procedures as described below.

30 The MIC does not handle biological attacks (House of Lords 2009a: Q25).

31 The Solidarity Clause may eventually change this. For an analysis, see Myrdal and Rhinard (2010).

32 Commission of the European Communities (2004). "Communication from the Commission to the European Parliament, the Council, the European Economic and Social Committee and the Committee of the Regions – Reinforcing the Civil Protection Capacity of the European Union." COM (2004) 200 final. Brussels: March 25, 2004.

33 CESIS stands for Common Emergency Communication and Information System.

34 European Commission (2002), "EU Focus on Civil Protection" Informational Brochure, Directorate General for the Environment, p. 5.

35 The participants in this network have their home affiliation in the ministries of Interior, Defence, and Foreign Affairs, in the emergency, rescue, and civil protection agencies, or in the permanent representations to the EU in Brussels.

36 The target group consists of technical, coordination, and assessment-orientated experts, and national personnel who work with civil protection modules. "Inventory of Crisis Management capacities in the European Commission and Community Agencies." Internal Commission Document. Brussels: July 31, 2009.

37 The exercises are primarily organized by Mechanism countries, although three "desk-top" exercises have been organized by the EU under the Council's Crisis Coordination Arrangements (House of Lords 2009a: 28–30).

38 A more detailed breakdown: training, €5 million; exercises, €2 million; modules, €2 million; preparedness projects, €1.35 million; prevention projects, €1.7 million; transport inside EU, €1.83 million; transport outside EU, €6.66 million; early warning systems, €1.13 million.

39 The data in this section was taken from a report prepared for our research project by Simon Hollis (2009). The sources of information come from personal correspondence with the MIC, official EU documentation, NATO situation reports, and press releases. It should be noted that there was some difficulty in obtaining a complete data set for the provision and type of resources provided through the MIC in the period 2002–7.

40 Member states can also request assistance from NATO's Euro-Atlantic Disaster Response and Coordination Center (EADRCC), which runs a similar program for NATO member states.

41 The case description is based on Ekengren (2007a).

42 Since the Czech Republic was a candidate country at the time, DG Enlargement was also involved.

43 This section is based on "Report on the loss of the tanker *Prestige* – In view of the public hearing of 19 March 2003 at the Committee of Regional Policy, Transport and Tourism of the European Parliament." Commission Staff Working Paper: SEC (2003) 351. Brussels: March 17, 2003.

44 The case description is based on Ekengren, Matzén, and Svantesson (2006: 49–56).

45 This case description draws from Ekengren, Matzén, and Svantesson (2006: 71–102).

46 The MIC also participated in the evacuation of EU citizens. It took part in the interconsular telephone conferences organized by the Presidency, which aimed to coordinate the evacuation efforts in Thailand.

47 European Commission (2004). "The European Commission coordinates EU civil protection support to catastrophe areas in South Asia." Press Release IP/04/1544. Brussels: December 31, 2004.

48 Communication from the Commission to the European Parliament and the Council on reinforcing the Union's Disaster Response Capacity. COM (2008) 130.
49 Special Eurobarometer 383, published in June 2012.
50 Communication from the Commission to the European Parliament and the Council on reinforcing the Union's Disaster Response Capacity. COM (2008) 130.
51 Communication from the Commission to the European Parliament and the Council on reinforcing the Union's Disaster Response Capacity. COM (2008) 130.
52 Commission of the European Communities (2005). Communication from the Commission to the Council, the European Parliament, the European Economic and Social Committee and the Committee of the Regions: "Reinforcing EU Disaster and Crisis Response in Third Countries." COM (2005) 153 final. Brussels: April 20, 2005.
53 Proposal for a Decision of the European Parliament and of the Council on a Union Civil Protection Mechanism. COM (2011) 934 final, p. 4. December 20, 2011.
54 Communication from the Commission to the Council, the European Parliament, the European Economic and Social Committee and the Committee of the Regions: "Reinforcing EU Disaster and Crisis Response in Third Countries." COM (2005) 153 final. Brussels: April 20, 2005.
55 Communication from the Commission to the European Parliament and the Council on reinforcing the Union's Disaster Response Capacity. COM (2008) 130.
56 Testimony of Bruce Manning (House of Lords, 2009a: Q46).
57 Official documents have increasingly emphasized the usefulness of the Mechanism in comparison to more traditional humanitarian aid instruments. See "Communication from the Commission to the Council, the European Parliament, the European Economic and Social Committee and the Committee of the Regions: Improving the Community Civil Protection Mechanism." COM (2005) 137 final. Brussels: April 20, 2005.
58 In his testimony, Mr. Das asserted that cooperation between the MIC and the EADRCC had improved recently: "There is a full and clear understanding between all Member States that in all emergencies which simultaneously involve the MIC and the EADRCC we will be sharing information on the assistance that we are providing and the actions that we are undertaking... That practical exchange of information has taken place already for several months (House of Lords 2009a: Q27)."

59 In a 2008 communication, the Commission evokes climate change as a reason to strengthen disaster response capacity. "Climate change and international security," Paper from the High Representative and the European Commission to the European Council. S113/08. In addition, the Commission has begun to develop an EU strategy for Disaster Risk Reduction in Developing Countries.

3 *The EU as global crisis manager*

How emerging tools shaped ambitious policy aims

With so many cooks in the kitchen, it is, in some ways, amazing that any broth is produced at all.

(Howorth and Menon 2009)

This is the European way of doing things: a comprehensive approach to crisis prevention and crisis management; a large and diversified tool box; a rapid response capacity; playing our role as a global actor.

(Javier Solana)[1]

The EU steps on the international podium

In January 2003, the EU embarked on its first crisis management mission abroad. The EU Police Mission in Bosnia and Herzegovina (EUPM) would last seven years and had a staff capacity of 540 at its peak. The EUPM aimed to establish and train a multiethnic police service and to assist local authorities in conducting large-scale crime investigations. By building and strengthening key institutions in the criminal justice sector, the mission sought to facilitate a return to normalcy in a war-torn region (Merlingen 2009).

In the summer of 2003, an EU military mission landed in the Congo. Militias had killed hundreds of civilians. Thousands sought protection from the UN mission (then dubbed "MONUC"). When the UN battalion could not protect the terrified population, the UN called on the EU to send forces. The EU mission (ARTEMIS) of approximately 2000 soldiers drove the militias out within weeks and handed control back to the United Nations.

On December 26, 2004, a tsunami obliterated coastal areas of South and Southeast Asia, killing tens of thousands. Through DG ECHO, the EU immediately began funding response operations carried out by the Red Cross, UNICEF, the World Health Organization (WHO), and Oxfam. The EU coordinated the assistance of member states through its Monitoring and Information Centre and sent experts, search and rescue crews, doctors, medical aid, tents, and water to the area. The interconsular crisis network was used to locate and repatriate EU citizens in the area.[2]

These foreign missions marked the beginning of an important development: since the beginning of this century, the EU has begun to use and further expand its military and civil crisis management capacities, applying them to a wide variety of international operations. The EU has provided aid to citizens and countries affected by the Asian tsunami and to earthquake-stricken populations in Iran, Turkey, Morocco, Haiti, and Japan. It has led military peacekeeping missions in the Former Republic of Macedonia and in Bosnia-Herzegovina. It has sent missions to Lebanon and the Congo, Aceh, and Georgia (to name but a few adversity-stricken countries). The EU flag, in short, can be seen in crisis spots around the world.[3]

The EU has created these capacities at an extraordinary pace (Duke 2000; Van Ham and Medvedev 2002; Cornish and Edwards 2005; Hansen 2006; Howorth 2007; Kirchner and Sperling 2008; Greco, Pirozzi, and Silvestri 2010). The development of these capacities does not flow from the original mission of the European Community nor is it the outcome of grand bargains between key member states. The emerging profile of an international crisis manager is the – more or less unintended – result of an intriguing policy history that is still evolving as we write this.

Three shifts have shaped this policy history (cf. Menon and Sedelmeier 2010).[4] The sudden implosion of a long-reigning security paradigm after the fall of the Berlin Wall was one such shift. The search for a common foreign policy, a united voice for Europe, is another. The creation of crisis and disaster management capacity aimed at the assistance of overwhelmed member states (discussed in Chapter 2), and humanitarian assistance capacity for non-members, is a third shift that shapes this policy history.

Together, these shifts created both a need and an opportunity to define the EU's profile in terms of a global crisis manager. The sudden demise in 1989 of Western Europe's long-standing security paradigm prompted a hurried effort to reconsider and redesign Europe's joint security apparatus. This gave rise to the Common Foreign and Security Policy (CFSP). A key instrument to furthering a common foreign policy was the adoption of the European Security and Defence

Policy (ESDP): Europe's effort to create military means in support of a common political voice.[5]

Subsequent crises abroad, however, demonstrated in a painful and often embarrassing way that such a common voice is hard to find (Hyde-Price 2004). The EU escaped from further international embarrassment and stasis by agreeing on a remarkable step forward: the EU began to supplement its freshly developed military means with other civilian and disaster management capacities to assist troubled regions across the world. Within a brief period of time, the EU began to mix military and civilian methods of crisis management. The EU's profile became one of managing humanitarian crises, post-conflict situations, and the aftermath of natural disasters through a unique mix of methods.

This rapid shift – from a fledgling European military power to a recognized global crisis manager – was more a matter of evolution than design: it emerged quite unintentionally as EU institutions tried to fill in the grand ambitions imposed on them.

This evolutionary process is far from completed. Several recent developments are likely to have an impact on the current state of affairs. The Lisbon Treaty boosted the position of the High Representative, who now commands the European External Action Service. The DG Humanitarian Aid and Civil Protection (ECHO) became the home of the EU's Monitoring and Information Centre (MIC) and created the European Emergency Response Centre (a merger of MIC and ECHO crisis centers); this should enhance the Commission's role in the support of disaster-suffering countries. In the future, the EU may well be capable of applying an enhanced crisis management toolkit in a widening array of hot spots.

This chapter tells the remarkable story of the EU's emergence as a global crisis manager.[6] It sketches the policy history, describes the available capacities, and discusses the ramifications of this evolution. In the concluding section, we briefly reflect on the importance of the EU's external crisis management capacity for its position on the world stage.

The EU as a global crisis manager: a brief policy history

The demise of a traditional security paradigm

The EU grew up without any sort of "hard" capacity to address crises occurring outside its borders. In theory, of course, the EU could –

and did – throw its (increasingly growing) economic weight around and apply "normative pressures" to influence events outside its borders (cf. Manners 2002).[7] Yet it lacked the traditional trappings of a global power, such as an army or a strong foreign policy apparatus. Foreign policy and security issues remained firmly within the domain of national governments – and their defense alliance (NATO).

In its early years, the budding European Community had briefly considered joint security as a common goal. The founding fathers discussed a European defense community, but member states chose NATO as their preferred instrument for a joint defense (Smith 2003). Most Western European states wanted to ensure America's security commitment to Europe.

The postwar security paradigm that united Western Europe and North America for decades was characterized by an unambiguous threat perception, deep consensus on defense and security means, and a historically unique level of legitimacy for these means. Major strategic choices were informed and constrained by the bipolar division of the world into blocs (NATO v. the Warsaw Pact), which were locked in confrontation over ideology, territory, and geostrategic interests (Jones 2007).[8]

The major tools in this paradigm were nuclear deterrence in combination with collective and territorial defense. Defensive power was defined in terms of military capabilities, and security was seen as a function of defensive power (Kennan 1947).

The collective defense clause of the North Atlantic Treaty (1949) became the lynchpin for transatlantic cooperation. Another key rule of the traditional paradigm dictated that NATO's operations would be geographically confined (Art. 6) to the territories of its member states. The object of protection was the territorial integrity of the allied states (Kaplan 1999). NATO's legitimacy was predicated on the widespread (though not universal) belief that Western European states could not stand alone against the Soviet threat. EU member states displayed little inclination to undermine or supplant their beneficial ties with the United States, nor did they seriously consider an alternative to NATO.[9]

The EU enters the security arena

The end of the Cold War in 1989 caused a sudden and deep decline in the perceived legitimacy of this traditional security arrangement (Hogan 1992; McInnes 1994; Lundestad 1998; Gaddis 2005). The demise of the Soviet Union, the fall of the Berlin Wall, and the reunification of Germany created a new security context for European states. Traditional bipolar tensions virtually disappeared and territorial defense became seemingly irrelevant. Believing that the United States would eventually withdraw its troops from Europe, and facing the inevitability of German unification, German prime minister Kohl and French president Mitterrand pushed for more European integration and the addition of an explicit security dimension for the EU (Jones 2007: 37).[10] Europe had to be able to defend itself. But against whom or what?

The search for an answer to this essential question played out against a backdrop of the EU's deepening and widening in the form of the economic and monetary union and the greatest enlargement process in EU history – from 12 to 15 and, eventually, to 27 member states (including many former adversaries).[11] This transformation greatly affected the reshaping of security thinking in Europe.

Before the EU could formulate an answer (let alone a strategy), the outbreak of civil war (in 1991) and atrocious violence in the Balkans created a crisis for the Union. This war did not pose the traditional military threat of invasion. It merely called for an effective intervention, normally supplied by one of the superpowers. Some European leaders saw an opportunity for the EU to assert itself as a security player; Luxembourg's foreign minister Poos famously declared in 1991 that this was "the hour of Europe, not the hour of the Americans." EU member states, however, failed to provide a coherent and, above all, effective response to the events in former Yugoslavia. This major embarrassment was further heightened when the United States – reluctant and annoyed – finally stepped in to bring the war to an end through the Dayton peace agreement (Daalder and O'Hanlon 2000a, 2000b; Christopher 2001).

The EU's failure strengthened a latent yet widespread desire to play a larger role – one befitting an economic superpower – on the world's political stage (Jones 2007). The EU had become an economic giant, but remained a political dwarf. The Balkan tragedy, unfolding on the doorsteps of the Union, demonstrated that this "mismatch" between political and economic capacities was no longer acceptable. If the Union were to play a role on the international stage, its members would have to learn to speak with one voice.

This ambition has a long history with very few successes. The first modest efforts to coordinate the foreign policies of the member states – known as European Political Cooperation (EPC) – dated back to the 1970s. These efforts yielded few results over the years beyond "declaratory" policies.[12] The fall of the Berlin Wall created new momentum. In 1992, a Common Foreign and Security Policy (CFSP) was adopted in the Maastricht Treaty. The aims of this common policy were sweepingly broad and understandably vague since they included both narrow self-interests (safeguarding the EU's fundamental interests and independence of the Union) and international ambitions (preserve peace, strengthen international security, promote international cooperation, and develop democracy).[13] Yet few instruments were adopted and limited authority was delegated to help bring about a coherent foreign policy.

The Petersberg tasks

The Amsterdam Treaty (1997) strengthened the CFSP by creating the position of a High Representative (Javier Solana became the first "face" of the EU's foreign policy). Importantly, the High Representative (HR) was equipped with a policy planning and early warning unit to monitor and assess potential threats.[14] The Amsterdam Treaty added more concrete international goals to the wording of Maastricht by including the so-called Petersberg tasks: humanitarian and rescue tasks, peacekeeping tasks, and tasks of combat in crisis management, including peacemaking.[15]

The Petersberg tasks had been conceived within the Western European Union (WEU) and were officially adopted in a declaration of June 19, 1992. The WEU at that point would be included in the EU to "develop WEU as the defense component of the European Union and as the means to strengthen the European pillar of the Atlantic alliance."[16] The emphasis on the Petersberg tasks helped ease the fears of many with regard to bringing a defense organization into the EU, which could jeopardize the status of NATO and military non-aligned members.[17]

By the end of the 1990s, there was a "substantial institutional framework" for the CFSP, which included earmarked budgets, offices, and procedures (Smith 2003: 564). In practice, however, the member states still did not speak with one voice on crucial international issues. New members and old sensitivities placed heavy constraints on a common foreign policy.

The 1997 election of UK prime minister Tony Blair generated a game-changing impulse. In December 1998, the Franco-British summit at St. Malo resulted in a declaration that the EU "must have the capacity for autonomous action, backed up by credible military forces" (Jones 2007: 85). France and the UK agreed that the EU needed its own military apparatus. A few months later, with the escalating events in Kosovo fresh in the mind, the EU Council in Cologne (June 1999) announced "a decisive step forward" and decided to "give the European Union the necessary means and capabilities to assume its [CFSP] responsibilities" by establishing the European Security and Defence Policy (ESDP) (Jones 2007).

The ESDP was to provide the military means to help perform the Petersberg tasks, which is "the area where a European capacity to act is required most urgently" (cf. O'Hanlon and Singer 2004).[18] Several institutions were created for institutional support (including the Political and Security Committee, the Military Committee, and the Military Staff).[19] The operational character of the ESDP thus triggered "the creation of a distinctive sub-set of institutions primarily charged with the planning and conduct of crisis management operations" (Grevi, Helly, and Keohane 2009: 19). This institutional subset would come to define the crisis management capacities of the EU.

A narrowing of the mission

In spite of the CFSP's creation, the member states found it hard to agree on a common foreign position in response to external crises. The Kosovo war in 1999 and the events of September 11, 2011, created acrimonious foreign policy challenges for the EU. The subsequent "War on Terror" and the Iraq invasion laid bare deep divisions within the EU. While the EU had begun to develop joint crisis management capacity, it could hardly agree on common positions or agree on when to deploy (military) crisis management missions.

In response to the deep divisions with regard to foreign policy goals (Bailes 2005; Cottey 2007: 85), the EU's High Representative Javier Solana sought to define the EU's security aims through the formulation of the so-called European Security Strategy (ESS) (2003).[20] This brief and remarkably well-written document suggests that the EU should play a role on the world stage within the context of an effective multilateral system (Ekengren 2007b). It reflects the EU's human security paradigm, which prescribes that the EU applies politics rather than guns in its handling of conflict situations (Buzan 1991; Waever *et al.*

1993). The ESS identifies five threat areas: weapons of mass destruction, terrorism, regional conflicts, state failure, and organized crime (disasters were not on the list).

The proposed policy focus chimed with the emerging consensus among Western states with regard to the importance of strengthening capacities for peacekeeping and humanitarian assistance (Smith 2003; Biscop 2004). The ESS gave the EU an opportunity to demonstrate "added value" in a specific niche that remained unclaimed by national powers.

While providing a rationale, the ESS did not provide a clear strategy for the use of the EU's budding crisis management capacity. The relation between the Petersberg tasks and the "hard" security and defense aims of the CFSP was tactfully ignored. As Hansen (2006: 49) observes, "the ESS offers little in the way of prioritization. Instead, it sketches a wide area of operation and leaves all doors open for potential future missions. There is very limited strategic thinking on when and where ESDP operations should be deployed and how each fits an overall set of aims."[21]

But the ESS did open the door for a rethink of the use of military tools. The newly developed ESDP tools were only to be used for a specific type of situation, embodied in the Petersberg tasks, which could count on widespread support.[22] But military tools clearly would not suffice for peacekeeping and state-building missions. It was soon accepted that the Petersberg tasks would also require *civilian* crisis management capacity, which the EU subsequently set out to develop.

Before the ESS saw the light of day, the first ESDP operation had started (January 2003). It was civilian: a police mission to Bosnia. The mission was not an easy one and not without flaws (Nowak 2006c: 15). But it started a process of discovering how the EU could effectively operate in international hot spots. Intriguingly, the first ESDP crisis management missions took place in the same year in which disagreement reigned over high politics regarding the war in Iraq. While many may have expected the de facto end of CFSP itself, the envisioned tools for CFSP – namely ESDP – began to shape the EU's profile as an international crisis manager.

The EU's activities in the international crisis arena today converge around a unique combination of military and civilian instruments (Biscop 2005; Howorth 2009; Ioannides 2010). As of this writing (summer 2012) there are 16 ongoing military and civilian ESDP/CSDP missions (12 have been completed) around the globe, ranging from border control assistance in Gaza to peacekeeping missions in the Democratic Republic of Congo (DRC) and police training programs in Kosovo and Afghanistan. The units range from legal advice teams consisting of a handful of experts to thousands of military troops. A complete list of ESDP/CSDP missions can be found in the appendix of this chapter.

Three pillars of external crisis management capacity

In what might be broadly called its "foreign crisis management missions," the EU can make use of three types of toolboxes. Two of these toolboxes – the military and civilian crisis management capacity – fall within the CSDP domain introduced above. These capacities are largely governed by the same EU institutions. The third, humanitarian assistance, is part of the Commission and is rarely connected to the CSDP domain (but it is becoming an increasingly important tool in the EU's crisis management missions). Together, these toolboxes enable the EU to address a wide variety of crises and disasters wherever they may occur. Let us now take a look at these toolboxes and see how they are organized.

The EU's military capacity

The EU does not have its own army, but it can bring together armed forces from the member states. Many conditions conspire against successful deployment of joint military troops in support of one European voice: all member states need to agree to a joint military mission, the EU requires international legitimization (preferably operating under a UN-adopted mandate), and the "host country" needs to accept the EU mission. Moreover, joint missions require a coordinated approach toward capacity building and maintenance in the member states (cf. Giegerich 2009).

The Helsinki Headline Goal (1999) set a first ambitious target: by 2003 member states would contribute toward the creation of a Rapid Reaction Force (60 000 troops to be deployed within 60 days, to stay at least one year in a crisis region, and to be self-sustaining).[23] But even in the wake of the 9/11 attacks, EU member states were very slow to deliver the promised goods. The Laeken summit (2001) produced the

European Capabilities Action Plan, which aimed to remedy shortfalls in available joint capacity.

Following NATO's 2002 shift to create a more flexible NATO response force, the EU decided to "go small" (Grevi and Keohane 2009: 73). The member states committed themselves "to be able to respond with rapid and decisive action" (Lindstrom 2007: 80). This was, of course, in line with the stated aims of the ESS to act rapidly in the face of crises across the world. To that effect, the newly revised Headline Goal 2010 announced the idea of EU Battle Groups: multinational units formed by a "framework nation" or a multinational coalition of member states, consisting of 1500 soldiers, deployable within 5–10 days, and able to stay 30–120 days.[24] The Battle Group idea compensated for the faltering progress in accomplishing the first Helsinki goals (Cornish and Edwards 2005: 805).

The EU's first military operation without direct NATO support (ARTEMIS) was a true test case for the independent deployment of Union capacity (Howorth 2007: 233–5). The task was to protect Congolese civilians, the airport of Bunia, refugee camps, and UN personnel (Ulriksen, Gourlay, and Mace 2004). In addition, the EU mission was to prepare the ground for an expanded UN mission later that fall. France offered to act as a framework nation for such an EU mission. The French forces were highly experienced in carrying out operations in Africa; many of the participating regiments had been in Africa before.[25]

Overall, ARTEMIS was regarded as a success (Gowan 2005; Ladzik 2009).[26] The mission drove the militias out of Bunia, which allowed 60 000 refugees to return home and humanitarian aid workers to do their work (Homan 2007). The French military command worked together quite effectively with the EU's Political Security Committee and the European Union Military Committee (Cornish and Edwards 2005: 808). In this first case of joint EU combat, there were no European losses (Ulriksen, Gourlay, and Mace 2004: 516–19).

This first autonomous military operation also unearthed some deficiencies, which helped to prepare for subsequent missions (Ulriksen, Gourlay, and Mace 2004; Grevi, Helly, and Keohane 2009).[27] The ARTEMIS mission provided a template for the new Battle Group units, which became operational in 2007. Member states agreed to strive toward EU targets in areas such as intelligence, computer networks operations, protection against roadside bombs, and increased availability of helicopters. But progress in capacity development has not been as quick as hoped for.[28] The financial crisis has made it much harder for member states to reach their targets (Greco, Pirozzi, and Silvestri 2010; Simon and Mattelaer 2011).[29]

The EU has not shown much willingness to engage in independent military adventures since. This brings us to an emerging paradox: while the EU has created an impressive capacity to organize and deploy armed forces across the world, the demand for that capacity has not grown. In fact, most of the EU missions abroad have been civilian crisis management missions. It is to this capacity we turn next.

The EU's civilian crisis management capacities

The EU has also developed non-military or civilian crisis management (CCM) capacity.[30] CCM refers to the "intervention by non-military personnel in a crisis that may be violent or non-violent, with the intention of preventing a further escalation of the crisis and facilitating its resolution."[31] This capacity was created as an afterthought to military capacity (mainly at the behest of Nordic and other "neutral" EU member states) and it developed in the military's shadow (Hansen 2006: 11; Giegerich 2008; Gebhard 2009).[32]

The 2000 Feira European Council approved four priority areas: police, rule of law, civilian administration, and civil protection. The number of tasks for civilian crisis management was quickly expanded to include judicial and penal reform, human rights, political affairs, security sector reform, mediation, disarmament, demobilization, and media policy. The CCM capacity thus had to serve many different goals.

The member states committed to making available specific numbers of national experts for civilian missions (the so-called Civilian Headline Goals).[33] This "rigorous attempt to identify personnel requirements for possible deployment scenarios" (Korski and Gowan 2009: 43) was followed by National Action Plans to help guide member states toward fulfillment of their respective Headline pledges.[34]

The EU honed its civilian crisis management capacities in the Balkans. The first civilian ESDP mission was the EU Police Mission in Bosnia-Herzegovina (January 2003). It sent an EU police mission (Proxima) to Macedonia. The EU Police Advisory Team, which followed Proxima, supported the government's police sector reforms.

In 2008, the EULEX mission started in Kosovo with the aim of supporting the strengthening of rule of law institutions and policing in that country.

The EU also became active further afield. The Georgia Rule of Law mission in 2004–5 (ten legal experts) involved technical assistance, humanitarian aid, food support, and rehabilitation programs in South Ossetia and Abkhazia. The EUJUST-Lex mission in Iraq trained senior officials of the Iraqi judiciary, police, and prison administrations. The purpose of the Aceh Monitoring Mission (AMM), including 130 EU monitors, was to oversee the peace agreement between the government of Indonesia and the Free Aceh Movement (GAM), which ended the civil war in 2005. In 2007 a police training mission was deployed to Afghanistan. Between 2007 and 2009 a military operation was carried out in Chad with a mandate to help the UN-protected refugees from Darfur. Teams have been sent to Palestine, the crossing point between Gaza and Egypt, and to the Moldova–Ukraine border.

The EU's CCM missions tend to be relatively small, emphasizing expertise over numbers. While it is easy to dismiss these missions as largely symbolic, the mere presence of EU CCM missions signals real commitment on the part of the EU. In the Balkans, for instance, the crisis management operations have been closely linked to other EU assistance tools and in this way functioned as a transition instrument from "stabilization to integration." They paved the way for reform of the juridical and security sector of these countries as a step toward full membership of the Union (Emerson and Gross 2007).

Civilian missions have made up the bulk of the CSDP operations. This should come as no surprise: these missions are perceived to work, tend not to unearth political tensions related to military missions, and fit the EU's "comprehensive security model" (Greco, Pirozzi, and Silvestri 2010: 5). Moreover, the civilian crisis management capacity comprises a varied toolkit, which can be used to address crisis situations very different in character, duration, and intensity (Ioannides 2010).

Organizing the missions: institutions and decision-making processes

In its continuing efforts to design institutions and processes for CSDP, the EU faces triple tensions that can run quite deep. When these tensions become manifest during a crisis, the political and administrative challenges mount quickly.

First, it is not always clear what type of event the EU faces: a military challenge, a civilian security situation, or a call for humanitarian assistance. When a disaster strikes in a vulnerable and violent area, a hybrid threat may emerge (bringing a combination of all three challenges). Such crises pose the question of which methods to use and which institution should be in charge of the operation. This is not always easy. While the military and civilian crisis management communities have been brought closer in recent years, the humanitarian assistance community is still organized separately (and falls under a Commissioner).

Second, the EU must negotiate the perennial tension between EU institutions and the member states. A mission requires careful deliberation (usually to achieve consensus between 27 member states) and speedy action (to stabilize a rapidly unfolding crisis).[35] To be effective and legitimate, the EU must combine consensus with rapid, effective decision making. This is one of the most profound challenges faced by any government facing a crisis or disaster (Rosenthal, Charles, and 't Hart 1989; Boin *et al.* 2005).

Third, the EU must negotiate inter-institutional tensions, especially between Commission and Council units. The organization of missions has traditionally been within the Council's remit, which has effectively limited the involvement of the Commission. Since its inception in 1999, the CSDP has grown into a veritable policy field consisting of a myriad of institutions and driven by complex decision-making processes.

In the decade since CSDP was initiated, the EU has built an intricate if complex institutional framework that can quickly and effectively negotiate these tensions.[36] This institutional framework is constantly evolving. The most recent impetus for change has been the Lisbon Treaty, which had significant consequences for the CSDP "pillar."

The Lisbon Treaty made the High Representative part of the Commission. The HR now presides over the EU Foreign Affairs Council and is at the same time Vice President of the Commission. The HR is also the head of the European External Action Service (EEAS), established as an autonomous EU body with the purpose of supporting all EU institutions in international matters. A raft of administrative bodies supports the organization and oversight of CSDP missions.

The European External Action Service

Many observers have tried to define the EEAS. Some see it as a foreign ministry in the making (Duke 2004). Others see a refinement of the intergovernmental status quo of CFSP – the "Brusselization" of national foreign policies (Wessels and Bopp 2008). Yet others hail it as the first supranational diplomatic service of its kind, reflecting the EU's commitment to smart power through coordination of hard and soft power (Cross 2011).

The stated purpose of the EEAS is to reinforce the political and strategic capability of EU foreign policy and crisis management. It should provide for continuity and improved institutional memory, compared to the earlier rotating presidency's often shortsighted objectives. A stronger overarching structure is thought to facilitate discussions between EU member states and within Union institutions about long-term foreign policy and security priorities – of great importance for the legitimacy of EU foreign policy in general and crisis management in particular.

The role of the High Representative and the EEAS is to coordinate the EU's external policies, instruments, and resources for a more coherent foreign policy. For this purpose, the EEAS was established as a merger of the external relations directorates in the Council Secretariat and the Commission. The EEAS is also a merger with regard to personnel: one-third of its officials is from, respectively, the Commission, the Council Secretariat, and the member states (in the form of seconded national diplomats).

The EEAS is closely linked to CSDP decision making by EU member states through the High Representative's presidency of FAC (Foreign Affairs Council) and the permanent chair of the Political and Security Committee (PSC) (see below). The HR chairs the "Group of Commissioners" that includes the Commissioners for Development, Enlargement and Neighbourhood, International Cooperation, Humanitarian Aid and Crisis Response, Trade, and Economic and Monetary Affairs.[37]

The EEAS Department for Crisis Response and Operational Coordination is set to play an important role when it comes to crisis management. Its managing director should provide a first assessment of an emerging crisis and is responsible for ensuring a coordinated response (coordinating the EEAS organs and relating their efforts to the longer-term work of the six other Managing Directors, who head the regional directorates).[38] This department also runs the EEAS Crisis Platform (which brings together all the players in a crisis) and houses the EU Situation Room and IntCen, which monitor worldwide events on a 24/7 basis.[39] The Consular Crisis Division is also placed in this department.

The EEAS's presence in the field through EU Delegations and EU Special Representatives should become a key resource for EU crisis management (the EAAS is present in 138 countries). The EEAS can build on these diplomatic networks to help prevent conflict and to support the management of acute crises. For example, the EU Delegation in the African Union (AU) "capital" Addis Ababa has supported the African building of capacity for conflict prevention, crisis management, and resolution (see further on page 70).

In contrast to the former Commission Representations, the Delegations are also responsible for political-security reporting and contacts – of great importance for risk and conflict assessment and CSDP missions. A close cooperation between the people posted on the ground and the ones sent out through missions is secured through the inclusion of political and military experts in the delegations' staff, in addition to the humanitarian aid experts who traditionally worked at Commission representations.[40]

The EU has worked with a growing number of special representatives and envoys to address tense situations in particular regions (Cornish and Edwards 2005; Hynek 2011). It has appointed special representatives to Moldova, Georgia, the South Caucasus, Central Asia, Kosovo, FYROM, Bosnia, Congo, Sudan, and the Palestinian territories. Some of these representatives – especially those in the Balkans – became quite powerful.

Other key institutions for EU crisis management

In CSDP matters, the most important body supporting the Council – the ultimate decision-making authority on CSDP missions – is the Political and Security Committee (PSC).[41] Consisting of ambassadors of the member states, this powerful body prepares the foreign ministers' meetings, oversees CSDP operations and may even receive authority to make decisions during an operation, which marks a "rare instance of explicit, albeit potential and circumscribed, delegation of decision-taking authority by the Council to a committee of national officials" (Grevi 2009: 30). The PSC has developed a distinctive working style – it remains intergovernmental – and has become truly institutionalized (Duke 2005; Howorth 2010). The PSC is supported in its work by the EU Military Committee (EUMC) and by the Committee for Civil Crisis Management (CIVCOM).

The EUMC directs the EU's military operations and provides the PSC with advice and recommendations on military matters, including strategic military options in a crisis situation, the operation concept and draft operation plan, and the termination options for an operation. It is composed of member state Chiefs of Defense (Gebhard 2009; Cross 2010). In a crisis situation, the EUMC exercises military direction and gives military instructions to the European Union Military Staff (EUMS).

The CIVCOM does the same for civilian crisis management: it oversees the management of civilian operations (Nowak 2006c: 23) and provides the PSC with information of available capacities. Contrary to the deeply experienced EUMS, CIVCOM consists of mid-career diplomats who come to Brussels with little experience in either civilian crisis management or EU affairs (Cross 2010).

Several bodies inside the EEAS provide operational support to the EUCM and CIVCOM: the European Military Staff (EUMS), the Civilian Planning and Conduct Capability (CPCC), and the Crisis Management and Planning Directorate (CMPD).[42] The EUMS provides recommendations and advice to the High Representative and performs the tasks of early warning, strategic planning, and situational assessment.[43] It houses the EU Operations Centre (first activated on March 23, 2012).[44] The EUMS is made up of military experts seconded by member states.

The CPCC (65 staff) plans, conducts, and reviews civilian CSDP operations under the political control and strategic direction of the Political and Security Committee and the overall authority of the HR. The CPCC was created in 2007 as a civilian counterpart to the EUMS and has taken on the structure of a civilian operational headquarters (Gebhard 2009; Simon and Mattelaer 2011).

The creation of the CMPD was initially put forward during the French Presidency in November 2008. The point of setting up this new structure was to "gather into a single body all the strategic planning capabilities until now spread out across the Council's structure . . . in order to conduct integrated strategic planning, including advance planning" (Simon 2010: 25). As such, the CMPD serves as an integrated strategic planning unit, located in the EEAS. According to Daniel Keohane (2011: 207), the creation of the CMPD marks an important step in the direction of better planning and implementation of CSDP missions.

A key planning and coordination role in the ramping up of joint military capacity is played by the European Defence Agency (EDA).[45] The EDA has assumed an active role in pushing the member states to deliver on the Helsinki Goals by suggesting specific national contributions (Cornish and Edwards 2005: 805).[46] The EDA produced "Capability Improvement Charts," which identified the progress or shortcomings of hundreds of "capabilities."[47]

In 2001, the Rapid Reaction Mechanism was established.[48] This somewhat misleadingly titled funding mechanism allows the Commission to rapidly disburse funds to help stabilize crisis situations. In 2007, it was aptly renamed the Instrument for Stability (IfS) and the annual budget was enlarged from €30 million to €220 million. The so-called Foreign Policy Instruments, which, in addition to IfS, includes the EU's financing of CFSP operations and election observation missions, falls under the responsibility of the High Representative. The practical management of these resources, however, remains under the authority of the Commission, which follows the guidelines established by EEAS.[49]

In sum, we can see that a dense institutional framework comprising civilian and military crisis management bodies has materialized in the brief period of a decade. The creation of the Civ-Mil cell and the subsequent creation of the Civilian Planning and Conduct Capability to provide a civilian counterpart to the EU Military Staff, combined with the setting up of the Crisis Management and Planning Directorate (merging the military and civilian working groups of the Council Secretariat) to integrate civilian and military planning at the strategic level, serve as good examples of this trend. The Lisbon Treaty and the creation of the European External Action Service have further brought the civilian and military bodies closer to one another.

CSDP and other international organizations

When the EU enters the international crisis domain, it usually cooperates with other international organizations (IOs). Over time, the

EU has developed close working relationships with a defined set of IOs: NATO, the African Union (AU), the UN, and (to a somewhat lesser degree) the Organization for Security and Cooperation in Europe (OSCE).

CSDP–NATO: evolving complementarity

Analysts have traditionally paid most attention to the EU's relationship with NATO.[50] The pronounced aim to formulate a common security strategy and create supporting capacities was seen by many as a first step toward the creation of a competitor to NATO.[51] The United States did not enthusiastically support the idea of independent EU military capacity. The EU clearly has not "drifted apart" as some feared. In fact, CSDP has taken ideas (and its first leader) from NATO, has forged close ties with NATO, and has carved out a distinct niche vis-à-vis NATO (Keohane 2009; Smith 2011).

NATO has had a fundamental impact on the CSDP's institutional design, policies, and experience (Cornish and Edwards 2005; Gänzle and Sens 2007; Keohane 2009; Varwick and Koops 2009: 120–1; Ojanen 2011; Simon and Mattelaer 2011). Most of the states giving birth to CSDP were NATO members and many of the founding key personnel either had experience in NATO or double-hatted in both organizations (Cross 2010). NATO members had followed the same guidelines for national defense planning for decades. When CSDP initiated its first military mission (Concordia in FYROM), NATO's Deputy SACEUR (Supreme Allied Commander Europe) functioned as operational commander and CSDP forces relied on NATO for planning and support (the working relation was positively evaluated). Much of the CSDP's activities in the Balkans has depended on cooperation with NATO (Cornish and Edwards 2005).

While an "inherent skepticism" governs perceptions of NATO in Brussels (Norheim-Martinsen 2009: 15), cooperative arrangements "on the ground" reportedly work well (Cornish and Edwards 2005; Howorth 2009; Smith 2011). These working relations are founded on the so-called "Berlin Plus" agreement (2003), which provides the Union access to NATO planning facilities and capabilities for EU military operations (Sperling 2010).[52] It laid down the political principles of the partnership: effective mutual consultation, equality, and decision-making autonomy for both entities.[53]

Meetings between the two organizations take place at different levels including that of foreign ministers, ambassadors, military representatives, and defense advisors. The North Atlantic Council and the Political and Security Committee meet bi-monthly.[54] In addition, there are regular staff contacts between NATO's International Staff and International Military Staff and the European Union's Council Secretariat and Military Staff as well as the European Defence Agency (Norheim-Martinsen 2009). The NATO–EU Capability Group ensures coherence and mutual reinforcement of NATO and EU capability development efforts. Permanent military liaison arrangements have been established to facilitate cooperation at the operational level. A NATO Permanent Liaison Team has been operating at the EU Military Staff since November 2005 and an EU Cell was set up at SHAPE (NATO's strategic command for operations in Mons, Belgium) in March 2006 (Norheim-Martinsen 2009). Since most member states belong to both the EU and the NATO process there are many channels for the two-way flow of norms and learning on lower administrative and expert levels.[55]

Close cooperation, especially on the ground, enabled the EU to develop its distinctive crisis management capacities, which are largely complementary to those of NATO (Cornish and Edwards 2005; Howorth 2009). The dysfunctional relationship between Turkey and Cyprus has prevented deeper cooperation (especially with regard to intelligence sharing). Yet, Europe continues to depend on NATO as a guarantor of its security, so sustained cooperation remains likely in the foreseeable future.

CSDP–African Union: cooperation through mentorship

The CSDP has forged a close cooperation with the African Union (AU), which has given rise to a substantial number of missions on the African continent. The AU adopted a "Common African Defense and Security Policy" (CADSP) in 2004, largely modeled on ESDP (Helly 2009). In its Strategy for Africa (2007), the EU pledged to support and enhance the AU's capacity for conflict prevention and crisis management. This should result in an "African Peace and Security Architecture," which will include the budding African Standby Force (ASF), a Continental Early Warning System and Security Sector Reform.[56] The "Action Plan for Africa" explains how the CSDP will use military and civilian instruments to help the AU plan and conduct peace support operations autonomously.

The AU has been involved in a range of peace support operations, including Rwanda (2003), Burundi (1993 and 2003), DRC (1999), and Ethiopia-Eritrea (2000). It has worked with CSDP in Sudan (AMIS 2004), the Central African Republic (FOMUC), and Comoros (AMISEC). Through the so-called Peace Facility (a €250 million fund established 2004), the EU funded many of these operations (except weapons and ammunitions).[57]

CSDP offers support in the field of training, technical assistance, and funding of AU personnel. It contributes with doctrine and standing operating procedures to the ASF. This includes participation of military and civilian expertise from the General Secretariat of the Council. The EU Military Staff is closely involved in joint workshops and the Secretariat has contributed to the development of concepts for police, rule of law, and civilian administration. The EU is funding many of these workshops and has a military liaison officer and police representative at the Commission delegation in Addis Ababa.

CSDP and the United Nations

In its short history, CSDP has also developed close cooperation with the United Nations (UN). It is the CSDP's policy to select and organize missions that are explicitly in accordance with the UN Charter (Gowan 2009). As many missions were organized either on request of the UN or with the blessing of the UN, the relationship with the UN has provided CSDP with the legitimacy that the EU's budding foreign project so badly needed (Grevi, Helly, and Keohane 2009: 403).[58]

At the same time, the UN needs CSDP to help fulfill some of its own ambitions (Ojanen 2011). The unique and comprehensive strategic approach to crisis management developed by CSDP has become a valued resource. One might say that the UN and CSDP are complementary organizations: CSDP assists the UN and operates under its umbrella, which provides the EU with part of the international action it craves; it sometimes helps the UN that the CSDP can go where the UN cannot go (Russia could not veto a CSDP mission to Georgia).

CSDP works closely with the UN's Department of Peacekeeping Operations (DPKO). Since 2003, a Steering Committee between EU and UN Council Secretariats has been active; the CSDP has had a military liaison with the UN since 2005. While there have been problems when CSDP and the UN operate military missions in the same theater (Gowan 2009), cooperation between CSDP and the UN is widely considered successful and likely to grow (Hynek 2011: 87).

CSDP and the OSCE

In the Cold War days, the Organization for Security and Cooperation in Europe (OSCE) was an important venue to discuss and negotiate contentious issues separating East and West. The fall of the iron curtain more or less deprived the OSCE of its *raison d'être* and some say CSDP has replaced it (Lynch 2009). But the OSCE still exists and the CSDP cooperates with it, especially when it deploys in Eastern Europe. Overall, the relation appears to be one of distant cousins and there seems little reason to suggest it will intensify in the future.

Adding to the EU's crisis management toolkit: humanitarian assistance

While civil protection was one of the stated aims of civilian crisis management, the creation of ESDP did not lead to independent capacity in this direction. The EU simply made use of – and still depends on – various instruments that already existed in the EU. We outline two here (and they are both in the Commission): DG ECHO's humanitarian aid capacity and the Civil Protection Mechanism. These instruments, as we noted in Chapter 2, were rarely used in concert. That might change since both instruments have been brought together in DG ECHO.[59] It is worth noting here that humanitarian assistance traditionally received the lion's share of EU funding for crisis management.[60]

Well before the ESDP came into existence, the EU responded to foreign disasters through the Commission's Humanitarian Aid department (ECHO), established in 1992.[61] ECHO's main mode of operation is through the funding of NGOs such as the Red Cross and international organizations such as the UN, which, in turn, provide direct emergency relief to victims of natural disasters or armed conflict. ECHO grants cover emergency aid, food aid, and aid to refugees and displaced persons worth a total of more than €700 million per year (in parity with US humanitarian aid).

The EU also has its Civil Protection Mechanism, as discussed in the previous chapter, which helps pool resources should a member state, overwhelmed by a natural disaster, request assistance. While initially aimed to facilitate mutual assistance within the Union, the Civil Protection Mechanism in recent years has become more of an instrument of support for the EU's disaster response around the world (Duke and Ojanen 2006).[62]

After the September 11, 2001, attacks, more than 1000 rescue workers from the member states were coordinated through the Mechanism for missions across the Atlantic (de Wijk 2004). The Mechanism was used when Morocco was hit by an earthquake (February 2004). In 2006 it was used to help member states evacuate their citizens from Lebanon and to send experts for the education of locals to clean up the oil spill caused by Israeli bombings. Other notable disasters during which the Mechanism was used include the earthquakes in Algeria (2003), Iran (2003/2004/2005), Pakistan (2005), Haiti (2010), and Japan (2011); the tsunami in South Asia (2004/2005); and Hurricanes Katrina and Rita in the USA (2005).

The response to such events falls under the Commission's remit, specifically under that of the Commissioner for International Cooperation, Humanitarian Aid, and Crisis Response (ECHO). She heads the European Emergency Response Centre (ERC), which coordinates the EU's Response Capacity.[63]

The Commissioner has operational authority, if co-approved by the rotating Presidency of the Council and the High Representative, to organize EU-flagged disaster response missions. Moreover, the Commission makes humanitarian aid decisions internally, with only advisory input from member states via an intergovernmental committee on Civil Protection (named "ProCiv"). In the future, it seems probable that the EEAS will also become increasingly involved in this area, which makes the issue of coordination between DG ECHO and the EEAS an important one. The EU mission to Haiti provided some valuable lessons in this regard.

One noteworthy aspect of the complex Haiti crisis was the pulling together of different "security communities" for disaster management: "the humanitarian aid community," "the civil protection community," and "the military and security community." This was at times a painful experience, challenging ingrained principles and prevailing mind-sets within different organizational cultures. The different actors within these communities at the EU level and within member states were not always keen to coordinate their actions – but to a certain extent they had no choice. Most of the tension arose between the traditional humanitarian perspectives, promoted mainly by the Commission, and the more political approach embodied by the emerging European External Action Service and the military structures.

But the Haiti mission also demonstrated that EU institutions on "different sides of the house" could cooperate. A temporary cell was established in Brussels, "EUCO Haiti," to coordinate member states' offers of military and security capacities. The cell was placed in SitCen (and, thus, closely linked to the HR within the emerging European External Action Service) and on the ground in Haiti.

This new coordination cell caused quite a stir in Brussels and confusion in capitals. The main reason was the perceived duplication between EUCO and the MIC. From the perspective of the European Commission, EUCO was seen as an attempt to create a parallel, "rival MIC structure" in the European External Action Service. For member states, the arrival of a second EU mechanism for coordination of military assistance was not necessarily considered an improvement (particularly for those member states that made little distinction between civilian and military capacities for disaster response). The establishment of EUCO called for a new set of national contact points within the military authorities and also the implementation of new procedures for the provision of assistance.

In order to defuse some of the criticism of this coordination cell, EUCO was repeatedly described as "coordination light" – implying a role limited to information sharing rather than more substantial resource coordination. Even if EUCO ended up as a rather "light" function, the initial ambitions were higher.[64]

Only days after the earthquake, the idea of an "EU Rapid Reaction Capacity" emerged. Vocal MEPs and certain member states argued that Haiti revealed the need for a stronger and more visible "European" capacity – beyond existing arrangements – to rapidly mobilize both civilian and military assets.[65] This call for a disaster force echoed earlier proposals (see for example the Barnier report 2006).

International cooperation

The EU is heavily dependent on international organizations to fulfill its disaster response ambitions. Whether it acts as a dispenser of aid money or as an organizer of member state presence in a disaster-stricken area, the EU must work with other international organizations. DG ECHO traditionally assisted stricken countries by funding the activities of such organizations as the United Nations High Commissioner for Refugees, UNICEF, Oxfam, the World Health Organization (WHO), the International Federation of Red Cross and Red Crescent Societies, and the United Nations Development Program. DG ECHO may set up a small field office, but the work is done by these organizations.

More recently, the EU has used its Civil Protection Mechanism to organize and send relief missions to disaster areas. This has brought the EU into a dense field of (sometimes hundreds of) IOs that often find it hard to coordinate efforts and compete for attention from media and funders (Comfort 1989; Nolte and Boenigk 2011). The EU has tried to address this problem by installing liaisons with other IOs.[66] The EU is a relative newcomer in this field and still has to find its bearings on the international disaster scene.

The EU has not cooperated much with NATO and its Euro-Atlantic Disaster Response Coordination Center (EADRCC), Rapid Response Forces and Combined Joint Task Forces (mobile headquarters) for international crisis management and humanitarian assistance. NATO has in recent years offered assistance during natural, technological, and humanitarian disasters (Cavanagh Hodge 2005; Rees 2006), but cooperation with DG ECHO appears to be limited at best.[67]

Analysis: how the EU's capacities became institutionalized

Twenty years ago, the EU had no crisis management capacities apart from DG ECHO (which mainly wrote checks to NGOs active in disaster areas). Then the EU initiated CFSP and, subsequently, ESDP. After a pioneering stage, we may well say that the various crisis management capacities have become rapidly institutionalized in the first decade after the EU seriously began to build capacities (Cornish and Edwards 2005; Gebhard 2009). This institutionalization curve can be mapped on two dimensions in particular.

First, the network of organizations that together govern the process of EU crisis management has become denser (and increasingly complex). This chapter has described a growing number of organizational units and organizations displaying intensifying relations between them and bringing together previously disconnected task areas.

Second, the EU's emerging identity as a global crisis manager has been received well, which suggests a growing degree of legitimacy. Member states appear to support the missions, as they allow the EU to participate in uncontroversial "UN-like" operations that do not intrude upon their sovereignty. Internationally, the EU has managed to gain support from the UN, NATO, and the "receiving" areas of EU attention.

The EU, in short, has come a long way. To explain the growing level of institutionalization of the EU's foreign crisis management capacity we make use of the institutional perspective laid out in Chapter 1.

Translating abstract political aims into practice: missions as laboratories

The EU has been likened to "a laboratory where options are explored for politics beyond the state, a toolbox for non-state-based governance, a pioneer in long-term interstate community building" (Nicolaidis 2004). The experimental nature of building the EU's toolbox is on fine display in the way CSDP has run its first missions.

There really was not much of a choice. The aims as set forth in the founding ESDP documents were highly ambitious but also ambiguous (Mérand 2008; Grevi 2009: 22). The ESS did not provide clear direction for the use of the EU's budding military capacity or its non-existing civil capacities; it merely offered a rationale for the use of military means to serve a set of goals that nobody could really disagree with. To be successful, the EU would have to formulate an effective and legitimate way of working. Our theoretical perspective predicts that the EU would discover how these aims are best achieved by modest experimenting.

This is exactly what the EU did in its first missions. The EU had never done military or civilian crisis management operations before. The decision-making system of the EU was not designed for rapid decision making (a prerequisite for any type of crisis management) (Wagner 2003). The EU did not have any planning capacity for such operations. All this had to be built from scratch. A solution was found through the use of experimentation and clever copying.

NATO's long experience with organizing large-scale operations provided an obvious source of inspiration and organizational templates. The first ESPD missions (operation Concordia in Macedonia 2003 and ALTHEA in Bosnia 2004) closely cooperated with NATO (a cooperation made possible by the Berlin Plus agreement). It was certainly helpful that Solana, as former NATO Secretary-General, could play such a crucial role in the development of the ESDP (Allen and Smith 2000: 102; Allen 2009).

The first missions were explicitly construed as "learning-by-doing exercises" (Allen and Smith 2005; Nowak 2006c: 15; Petrov 2010). Civilian missions were "policy laboratories" with "personalities mattering more than procedures" (Simon and Mattelaer 2011: 19). The EUPM training mission in Bosnia and the ARTEMIS mission in Congo both created templates that would be used to plan future missions (Korski and Gowan 2009).[68]

Effective practices had to be discovered the hard way. Keen to score an early success, the EU reportedly underestimated both the size and complexity of the task in Bosnia (Nowak 2006c: 26). Helly (2006: 99) argues the mission in Macedonia was "too ambitious and too difficult to implement." In her analysis of the Proxima mission, Ioannides (2006) shows just how ambitious (and, in a way, naïve) these police operations are in practice. The idea that a brief and necessarily limited operation can "reform" police forces flies in the face of decades of research findings that explain why reform under the best of circumstances is hard.

Experiments also helped to understand certain limits to the CSDP ambitions. Consider the Battle Groups (Gowan 2005). Created in reaction to "lessons learned" of early ESDP missions (Lindström 2007), this new military instrument has yet to be employed. The Battle Group concept encountered a familiar problem: a lack of consensus on the EU's role in the world (the explicit connotation with "war" makes it harder to reach the broad consensus that is required for an EU mission). It is no surprise, then, that the Battle Groups are the subject of ongoing debates on how to increase their "flexibility" and "usability."[69] In the civil war in Libya in 2011, a Battle Group was not deployed to enforce an arms embargo against the Gaddafi regime – a clear-cut case for which EU naval patrols would have been available and appropriate (Menon 2011). The feedback suggests that Battle Groups may be one step too far in the development of CSDP.

Emerging problems: limited budgets and coordination problems

The initial missions were ambitious in nature. There was political pressure to succeed and some missions were defined as successes even as they were launched (Hansen 2006: 20). But the missions displayed recurring and well-documented weaknesses (for summaries of the missions see Grevi and Keohane 2009; Korski and Gowan 2009). Two of the most fundamental weaknesses were budget constraints and coordination problems. These weaknesses, in hindsight, could have sunk EU ambitions to send uniformed missions abroad in the name of a common foreign policy.

A key problem has been the lack of resources to carry out missions (Nowak 2006c: 37). Even though the EU collectively ranks as the largest defense spender in the world after the United States, it has found it hard to staff its missions and provide them with the material needed to be effective. For instance, it took six months to find 16 helicopters and ten short-range transport planes for the Chad mission (Grevi and Keohane 2009: 69–70). All reports mention a lack of qualified personnel (for the civilian missions) and a lack of military equipment (especially transportation planes). Hansen (2006: 23) found that the operational command structure lacked staff, a significant operations center for military affairs, and a clear chain of command.

The principle of "costs lie where they fall" means that member states willing to contribute forces for an EU mission also pick up the tab.[70] The Commission pays some costs in the case of civilian crisis management missions, whereby different funds are governed by different DGs, all with their own criteria and decision-making procedures (Gourlay 2006b: 106).[71] The CSDP can benefit from the Instrument for Stability, but the demand for EU missions still outstrips available capacity and budgets.

A second weakness has been the dense institutional structure, which has made the chain of command (from the Council to the field) a long and opaque one.[72] Jean-Marie Guehenno, the former UN Under-Secretary-General for Peacekeeping Operations, has called attention to the "excessively complex institutional arrangements" that govern CSDP missions (Korski and Gowan 2009: 7).[73] Korski and Gowan claim that "Brussels overlords are happy to ignore advice from the ground" and tend to "focus on trivialities." As a result, they claim, directions from Brussels are often "unactionable fluff" (Korski and Gowan (2009: 56, 57).

In Bosnia, to give another example, the EU was represented by no less than five bodies: the EUSR, a monitoring mission, a police mission, a military operation, and the Commission delegation (Korski and Gowan 2009: 55). An operational "gap" can, thus, separate Brussels headquarters from "the field" (Hansen 2006: 49) – it does not require a military background to understand that such gaps are not beneficial to the effectiveness of the operation.

Reports on earlier ESDP missions are punctuated by observations of bureau-political tensions. A lack of coordination existed between ALTHEA and EUPM (the latter mission was micromanaged by the PSC, says Hansen [2006: 19]). The Proxima mission saw "battles of competence." Ioannides (2006) found rivalry on the ground and in Brussels. Helly (2006: 94) observed a "clash of personalities" and "inter-institutional and individual tensions." The Georgia mission was purely civilian but completely prepared by the directorate for defense issues (rather than the directorate for civilian crisis management) (Korski and Gowan 2009: 60). Guehenno (2009) advised Brussels to stop micromanaging operations and bring the civil and military rooms of the EU's crisis management house together (this has been partially accomplished since).

In addition, tensions have been reported between the Commission and Council. The political decision-making process was often divorced from deliberations in the Commission because PSC ambassadors did not always have a complete overview of Commission capacities (Gourlay 2006b: 104; Nowak 2006c). This has resulted in overlap. Two such examples are the overlap between the Council's Civilian Response Teams and the Commission's Advanced Planning Teams (Gourlay 2006b: 113; Nowak 2006c: 32; Grevi 2009: 105) and between the SitCen (now located inside the EEAS) and the MIC (Gourlay 2006b: 114; Khol 2006).

These tensions are fed by a difference in legal competences, which are not always as clear as they may seem. CSDP missions were supposed to focus on emergency situations requiring a rapid and dynamic response. The Commission would then take care of the next phase ("development"). In practice, things have been far messier (Korski and Gowan 2009: 53). The Commission even took the Council to the European Court of Justice when a proposed CSDP mission was perceived to infringe on traditional Commission territory (Gourlay 2006b: 121; Grevi 2009: 50).[74]

The capacity to adapt (and embed what works)

An important explanation for the EU's success (in terms of institutionalizing crisis management capacity) is the EU's ability to adapt. In the face of various criticisms and observed shortcomings, the EU has adapted its processes and structures. The EU's crisis management capacity as it exists today is largely the result of "learning by doing."[75] The EU has demonstrated a steep learning curve, which has allowed it to organize a variety of missions that are often run simultaneously (Cornish and Edwards 2005; Hynek 2011).

One example is found in the response to the capacities deficit through the so-called Headline Goals. In order to prod the member states to provide resources, the Union established the Helsinki Headline Goals, declaring that member states by 2003 should provide for 60 000 troops that could be deployed within 60 days, stay at least one year in the crisis region, and be self-sustaining. The objectives were soon revised in the aftermath of 9/11 and the Iraq war of 2003 (Giegerich 2008). The EU saw the need for new types of military operations and extended the list of tasks with disarmament, military advice and assistance, post-conflict stabilization, conflict prevention, and the fight against terrorism including third-country cooperation. When the ESDP shifted toward the use of civilian methods, the EU created Headline Goals to create civil capacity.

In response to the many coordination problems marring the missions, the EU introduced a host of new committees and units. A cursory overview since 2005 includes much-touted improvements such as the Civ-Mil Cell, SIAC, WKC, MAP, CPCC, APTs, CMPD, and the Civilian Operations Commander (Grevi 2009: 42ff.).[76] The Crisis Management and Planning Directorate (a merger of DG E directorates) was created (in 2008) to integrate civilian and military planning at the strategic level (Gebhard 2009). In a highly symbolic move, the Commissioner of International Cooperation, Humanitarian Aid and Crisis Response, Kristalina Georgieva, merged the crisis rooms of ECHO and MIC into the European Emergency Response Center.[77] The EU created Crisis Response Coordinating Teams, which act as "institution bridge-builders" (consisting of senior officials from the Council

Secretariat and the Commission who together prepare the overall Crisis Management Concept) (Norheim-Martinsen 2009).[78]

The EU also learned from its humanitarian missions. After the operation in response to the 2004 tsunami disaster, the Commission increased and improved training and exercises of its experts. To improve its assessment capacity "on the ground," the Commission enlarged its assessment teams (now teams of about five people are sent to disaster locations; previously one expert was sent). Moreover, an increase in the number of field-based ECHO humanitarian experts has been adopted. The coordination of EC humanitarian aid and member state bilateral aid has been enhanced through more standardized information flows. To realize this goal, a network of member states' humanitarian emergency focal points has been established.[79] Finally, the difficulties in expatriating EU civilians from Asia led to increased consular cooperation between the member states and an EU 24/7 information center for consular matters.

The EU's response to the Haiti earthquake in January 2010 presented an immediate test case for the post-Lisbon structure. The EU demonstrated that it had learned some lessons. The EU was able to deliver "boots on the ground" within a reasonable time frame (a better performance than its response to the Asian tsunami). The EU also sent 323 military police from France, Italy, and Spain, the first use of the European Gendarmerie Force.[80] EU institutions were able to rapidly gather and interpret the information needed to support coordination and to steer the flow of EU funding for humanitarian aid. The MIC benefited from close cooperation with DG ECHO (producing the joint field SitReps) and the integration with the UNDAC team on the ground. SitCen contributed analysis reports based on consular reporting.

The biggest lesson learned, however, pertains to the nature of successful missions. The EU has learned that effective interventions require a mixture of civil and military instruments, which, in turn, demand specifically designed supporting structures. If one compares the current bureaucratic structure with the one that existed five years ago, it is clear that the EU is able to adapt rapidly and with rigor.

One of the aims of the Lisbon Treaty and the reforms in the area of external relations was to allow a simpler answer to the question: Who speaks for Europe in an external crisis? The post-Lisbon structure indicates the EU's desire to improve civilian–military coordination and cooperation with the CSDP bodies now located within the new EEAS structure. The new "double-hatting" of the High Representative should help to address inter-institutional tensions.

The response to Haiti demonstrated that the wish for a single "EU face" (and a single phone line to Europe) was not solved with the appointment of the new High Representative.[81] The EU was still in the midst of "post-Lisbon turmoil." The new Lisbon Treaty, barely in place, had shifted units and created new institutions. Not all members of the new Commission had been confirmed.

As the head of the new European External Action Service, Baroness Ashton was given prime responsibility for the Union's response to the tragedy. But the new permanent President of the European Council, Herman van Rompuy, was the first among the high-level representatives from EU institutions to issue a public statement on Haiti, deploring the tragedy and calling for a resolute European response.[82] The first "EU face" to visit Haiti (more than a week after the disaster), was the outgoing Commissioner for Development and Humanitarian Aid, Karel de Gucht.

The scale of the disaster and its far-off geographical location presented many obvious problems. But the main challenge for Ashton had more to do with managing critical voices in Brussels and in capitals pointing to a lack of "EU visibility" and the need for a faster and stronger "EU response."[83] Ashton's decision not to fly to the site of disaster was widely considered a strategic communication failure.[84] Criticism was particularly hard from some member states, such as France, Poland, and Spain, who argued that Europe's foreign policy chief should have used her new position more efficiently to "wave the flag for Europe."[85]

The EU, in short, has proven quite adept at institutionalizing ad hoc arrangements that were initially intended to relieve coordination burdens and develop civil capacities (Gebhard 2009). The embrace of civilian crisis management methods triggered "the creation of a distinctive sub-set of institutions primarily charged with the planning and conduct of crisis management operations" (Grevi, Helly, and Keohane 2009: 19). Commission institutions such as DG ECHO and the MIC would become increasingly involved in this subset.

The missions displayed coordination problems, which were addressed by the design of, or experimentation with, new coordination structures. The various planning units and coordinating bodies have been formalized, enlarged, and embedded in the Lisbon Treaty. One might say that the European "art of coordination" has been firmly institutionalized within the EU crisis management domain.

Furthering legitimacy: finding the right niche

The EU's crisis management operations could not persist if member states did not consider them legitimate and fit for purpose. A first step to creating legitimacy for the missions was to consistently emphasize a close connection between the missions and the EU's long-standing tradition of creating peace and furthering good governance. While there was always a danger that certain missions would be seen as serving the interest of particular member states, the High Representative managed to maintain the perception that the missions served appropriate aims.

In selecting the missions, the EU played it wisely. The Union selected "crises" that did not pose real threats to the Union (or member states) and did not pose immediate danger to the EU nationals who were sent abroad. By working closely with the UN and NATO, CSDP missions drew legitimacy from these partnering IOs.

The missions worked well for the EU: they were small and clearly mandated. Critics charged that the EU missions were too modest in size and scope to make a difference. But troops returned without scandals, overt failures, or loss of EU lives. The fact that the EU had been able to deploy and run simultaneously different types of operations in different parts of the world demonstrated significant progress made since the EU launched its first missions (Cornish and Edwards 2005; Nowak 2006a; Grevi 2009).

We might conclude – with some exaggeration – that the world has accepted the emergence of the EU as an international crisis manager (cf. O'Hanlon and Singer 2004). The instruments originally intended to provide stability in the neighborhood are now requested and made available around the world. This has led to new requests for EU intervention, for more types of crises. It has also led to a broadening of the task environment, increasingly including complex disasters and demanding civilian–military missions. All this corresponds nicely with survey findings that suggest Europeans overwhelmingly support an EU role in the tackling of global threats (but not solely by military measures).[86]

The myth created out of the performance record – the EU as global crisis manager – may be out of proportion with the actual performance "on the ground." But it bridges "summit ambitions" and member state interests: most member state parliaments have very little appetite for military adventures in faraway conflict zones. It has thus made possible what seemed impossible not so long ago: to unite member states around a common foreign policy tool. It appears that the EU has begun to create real expectations. After the Haiti earthquake, the EU was criticized within days for not doing enough (while it mounted the biggest and best operation thus far).

Discussion: the role of leadership (or how Solana and his team created room for experimentation)

It is widely recognized that High Representative Javier Solana played an important role in the development and innovation of the CFSP (Smith 2003; Edwards 2005; Nuttall 2005) and a crucial role in the emergence of the EU's crisis management capacities. Solana played the role of top diplomat and policy entrepreneur: formulating policies and sanctioning practices that work, while selling them to the member states as the right way to accomplish the CSDP's aims.

Even though there was a large consensus behind the creation of a High Representative for CFSP in 1999 as a remedy for the weak performance in the Balkan conflicts, many EU member states and the EU Commission were worried that the new post would threaten their own powers. The position was therefore endowed with weak institutional powers and few resources. The Amsterdam Treaty limited the legal powers of the HR to "giving assistance" to the rotating EU Presidency, "representing the Council on political dialogue with third parties at the request of the Presidency" (Art. 26 TEU).

In practice, Solana managed to shape the role, organs, and political responsibilities of the position in those first years. The result was that the institutional structure developed ad hoc through an interplay between CSDP missions and what Solana deemed necessary. Having decided on new operations it became difficult for member states to put constraints on institutional innovations that for "practical" purposes were required for success, not least the ones that could help minimize the loss of lives. Solana thus facilitated the institutionalization of ESDP through experimentation and learning by doing rather than by imposing high-flying visions of "the Union's role in the world."

Solana was particularly effective in offering a rationale for CSDP that provided guidance for action while creating consensus among member states. To this effect, Solana produced the European Security Strategy (ESS) in 2003. The Iraq invasion had laid bare deep divisions within the EU. While the EU had begun to develop joint military capacity, it could not reach agreement on the most basic foreign policy issues. The ESS offered a narrative that facilitated joint action.[87]

Solana made sure that this budding "strategy" would be first implemented through a set of "easy" missions, which would allow the EU to develop well-anchored practices. Solana, in other words, created room for experimentation.

Solana recognized the Congo 2003 situation as a candidate for an ESS-inspired mission, not least because the notion of a European Security Strategy was under development at that time.[88] The High Representative cooperated closely with UN Secretary-General Kofi Annan in the process that led up to the UN's request for EU help in Congo (Hendrickson, Strand, and Raney 2007). Solana was given a wide mandate to play the role as Europe's diplomatic leader and sound out the practical preconditions for the operation. He sent his assistant, Aldo Ajello, to contact Uganda, Rwanda, and the Democratic Republic of Congo. He traveled extensively to meet African leaders and he played a crucial role as the EU's spokesman in the UN and as Annan's "diplomatic surrogate" to the EU (Hendrickson, Strand, and Raney 2007: 7–8).

Solana effectively assumed the role of institutional architect. The HR carefully listened to the member states and convinced them that a delegation of tasks to him would not jeopardize their interests, only bring efficiency in the implementation of their decisions. Solana also made sure the missions remained aligned with member state interests.[89] He stressed the importance of reforms to enable "crisis-mode" EU decision making (Barros-Garcia 2007: 5). While international relations theorists usually accord a minimal role for formal leaders in international organizations, we must conclude that Solana's influence on the institutionalization trajectory of CSDP has been significant and greater than member states would have predicted.

Conclusion: the inadvertent emergence of unique security capacities

The EU today is able to make a significant contribution to complex crisis management operations due to the broad range of political, economic, civilian, and military instruments at its disposal (Giegerich 2008; Howorth 2009; Ioannides 2010). The EU applies these instruments in a wide variety of arenas, ranging from state-building assistance in northern Africa to protecting ships from Somali pirates, from Pakistani floods to repatriating EU citizens from Lebanon.

The EU is, thus, fulfilling the long-standing ambition to play a role on the world stage. It does not do this through its original ambitions of projecting military force but by both delivering civilian resources useful for rebuilding states and dampening conflict (in the case of CSDP) and offering neutral, technical, and long-term assistance to countries and regions affected by crisis or disaster (in the case of humanitarian assistance).

The EU has successfully institutionalized these various crisis management tools. It has crafted an intricate network of organizations and policies that can respond to crises in a surprisingly fast manner. The tools have become tacitly accepted by member states and have become increasingly used (Giegerich 2008). The significant increase in security cooperation between member states has been called "one of the most striking developments in international politics today" (Jones 2007: 3; cf. Hansen 2006: 47). All this happened while the EU experienced a "big bang" enlargement from 15 to 27 members.

Looking back on a decade of CSDP in 2009, Solana described the development of the CSDP's crisis management capacity as "crucial to contributing effectively to international peace and security. It is the missing link. EU foreign policy used to be about declarations. Now the EU puts people in large, visible numbers on the ground and takes risks for peace" (Grevi 2009: 16).

This development must have come as a surprise to some academics. After the fall of the Berlin Wall, it was widely proclaimed that the old security paradigm was dead; a return to national security concerns was expected (Mearsheimer 1990; Waltz 1993; Calleo 2001). But European states did not turn back to national solutions (Coopers 2000, 2003). They pursued crisis management and closer military cooperation. And they did so in a new institutional constellation. Rather than using NATO for out-of-area operations, the EU created its own capacities to assist in international hot spots.

Striking as all this may be, this development is not the result of far-sighted design or visionary leadership of large member states (cf. Howorth 2010). Institutionalization of CSDP capacities has been the result of a depoliticized, functional, and problem-based approach: acting on requests, offering concrete help, and demonstrating competence. It was the inadvertent result of what organization theorists call experimentation and "mission creep." A tool in search of a mission, CSDP developed from an "afterthought" to the EU's most important foreign policy tool (cf. Hansen 2006). It widened and came to define CFSP (Wright and Auvinen 2009: 118). Dormant capacities in the EU's civil protection and humanitarian policy areas further enabled the development of the EU's emerging profile as a global crisis manager.

Leadership has been important in guiding and facilitating this process of institutionalization (cf. Selznick 1957; Boin 2001). But it would be simplistic to sketch the current capacities as the outcome of a well-implemented blueprint. The development of this capacity followed a functionalist logic, allowing tools to determine the mission in a depoliticized, "path of least resistance" manner.

Today, the EU is better known for providing civilian crisis management and "soft security" disaster response than anyone, especially politicians with visions of the EU as a military security actor, would have predicted ten years ago. Critics charge that the EU missions were modest in size and scope. That may well be true, but the point is that the EU is now conducting missions that were unthinkable 15 years ago. Moreover, the EU has demonstrated a steep learning curve and has developed a type of crisis management capacity that appears to work. This performance is cementing the EU's role as a global crisis manager and creating a demand for its services.

The EU will have to continue developing these capacities if it is to meet rising expectations (Cornish and Edwards 2005). That will require answering some fundamental questions concerning Europe's security aims and ambitions (Hyde-Price 2004; Boin and Ekengren 2009; cf. Beck 2005). Does the EU want to play the role of the global policeman (Wright and Auvinen 2009: 126)?[90] If so, how does this role of global crisis manager benefit the EU's security? If a mission is intended to "send a clear political signal" to a recipient country about the EU's full support of reforms and commitments to democratic values

(Helly 2006: 91), how does this benefit the EU's collective security? Does "penitentiary and probation service reform" in faraway countries really make the EU more secure? Moreover, what does the EU offer in response to new threats such as energy and climate change?

The EU also will have to consider whether its organizational structure and processes will allow it to accomplish its ambitions. We have documented a steady development and adaptation of the EU's security apparatus. But will that be enough in light of future challenges? Will the latest institutional innovations – especially the EEAS – enable the EU to act speedily, decisively, and effectively? Will the EU's focus on civilian crisis management suffice in the face of military threats (Simon and Mattelaer 2011)? What type of relationship between the EU and other security-oriented organizations such as NATO is needed in the future? We will return to these questions in the final chapter.

Appendix: CSDP Missions (2003–2012)[91]

EUPM/BiH (Bosnia, 2003): The first ESDP mission ("trailblazer and guinea pig") is a police mission. It is relatively large (540 staff at its peak) and will last seven years. It is considered fairly successful given the difficult circumstances.

Concordia (FYROM, 2003): The first military mission lasts eight months and is a takeover of a NATO mission (NATO's Deputy SACEUR stays on as EU operational commander). The EU relies on NATO for planning and support. The mission serves as a template for future missions.

ARTEMIS (Congo, 2003): A UN-supported operation, carried out by France (without NATO involvement). "A remarkable positive experiment in cooperation with the UN."

EUPOL Proxima/EUPAT (FYROM, 2003): Successor of Concordia, the ESDP's second police mission.

EUJUST THEMIS (Georgia, 2004): First rule-of-law mission, and a small one. Ambitious and marked by inter-institutional rife. This type of mission was originally seen as stretching the notion of civilian ESDP as Georgia was not even in crisis.

EUFOR ALTHEA (Bosnia, 2004): Largest mission to date (replacing NATO under Berlin Plus arrangements, similar structure as Concordia), considered successful. It coincided and conflicted with EUPM.

EUPOL Kinshasa/(DRC, 2005): Very small mission to DRC, lasting two years. Coordination with UNPOL remains wanting.

EUJUST LEX (Iraq, 2005): A small mission, big on symbolism. More useful for the EU ("therapy") than it was for Iraq.

EUSEC RD Congo (DRC, 2005): The first civilian mission to provide military advice ("experimental laboratory"). Marked by an evolving mandate, EU squabbling, competition with UN MONUC and China. Nevertheless, considered quite successful.

Support to AMIS and AMISOM (Darfur, 2005): Support mission to the AU. A very small mission, high on symbolism (peacekeeping training for Africans who arguably have more experience than their European trainers). It is considered important to develop the relation with AU.

AAM (Aceh, Indonesia, 2005): Considered a very successful peace-monitoring mission. Good collaboration with ASEAN countries but marred by funding problems, lack of MS unity, and Commission–Council strife.

EUBAM Moldova-Ukraine (Moldova-Ukraine, 2005): Providing border assistance managed by the Commission and in close cooperation with UNDP, this mission was considered a success.

EUPOL COPPS (Palestinian territories, 2005): Not effective, but symbolically important; the EU can do what the United States cannot do.

EUBAM Rafah (Palestinian territories, 2005): The customs assistance mission never had a chance to take off (it has been suspended since 2007).

EUFOR RD Congo (DRC, 2006): Relatively brief but large mission (2400) in support of UN. Cooperation with MONUC is "challenging."

EUPOL Afghanistan (Afghanistan, 2007): In a dense international environment, this mission operates without a NATO–EU agreement (in close cooperation with UNDP).

EUFOR Chad/RCA (Chad and Central African Republic, 2008): Logistically complex operation and politically somewhat controversial (France pushes hard for this mission, but the Commission is less than cooperative). French personnel (3700) operate with UN support.

EULEX Kosovo (Kosovo, 2008): Large rule-of-law civil mission (1642 personnel, including US personnel).

EU SSR Guinea-Bissau (Guinea-Bissau, 2008): Pushed by the Portuguese with little support from other member states. The mission cannot accomplish much due to civil war.

EUMM Georgia (Georgia, 2008): Highly successful monitoring intervention in close cooperation with OSCE.

EU NAVFOR (Somalia, 2008): Naval laboratory with unique collaborations, aimed at protecting trade vessels. Mission operates in close cooperation with NATO.

EUTM Somalia (Somalia, 2010): Training mission for Somali security forces based in Uganda. At its peak, 124 troops served in the mission.

EUAVSEC (South Sudan, 2012): Civilian mission aimed at strengthening aviation security at the Juba international airport in South Sudan.

EUCAP NESTOR (Horn of Africa, 2012): Training mission aimed at strengthening regional maritime capacities and rule of law in Somalia (complementing the missions Atalanta and EUTM).

EUCAP SAHEL (Niger, 2012): Training mission to assist Nigerian security forces fight terrorism and organized crime.

Notes

1 J. Solana. Address to the European Parliament on the EU Common Security and Defence Policy. Council of the European Union, S045/09, Brussels: February 18, 2009.

2 For an extensive case description, see Ekengren, Matzén, and Svantesson (2006, chapter 6).

3 Cynics remark that "big states fly combat missions where the EU flies the flag" (Smith 2011: 33).

4 These shifts are mutually interdependent to a degree but have largely autonomous drivers.

5 The Lisbon Treaty has changed the name to Common Security and Defence Policy. In this chapter, we continue the use of ESDP when we look back in time.

6 This chapter does not discuss, at least not thoroughly, the underlying motivations for selecting particular crises to act on. Why the EU acts on certain crises and not on others is a hotly debated issue among EU theorists. The neorealist school maintains an intergovernmentalist perspective, which holds that member states only act on a certain issue (crisis) if it is in their interest to do so (Hyde-Price 2007). The constructivist school, on the other hand, views the EU as a normative power seeking to spread its values wide and far (Manners 2008).

7 Over time, as membership became increasingly sought after, the EU developed a policy of "government by conditionality" through which it actively imposed EU policies on prospective member states (Lavenex and Wichmann 2009). This is an example of "hard" common foreign policy without military means. During the financial crisis that started in 2008, southern member states have been subjected to a similar form of "conditioned assistance" – reform in exchange for financial support.

8 We are skating here over France's relation with NATO, which has always been a bit temperamental (to say the least). For the story at hand, it suffices to say that France's commitment to a joint defense against the Russian threat has never been in doubt.

9 We should acknowledge here the Western European Union (WEU), an alliance of ten West European states. The alliance was formed during the Cold War as a collective defense pact, without including the United States. According to the treaty, it was founded "to create in Western Europe a firm basis for European economic recovery; to afford assistance to each other in resisting any policy of aggression; to promote the unity and encourage the progressive integration of Europe." The existence of NATO, the lack of US involvement, and the UK's ambivalence to the WEU ensured that it played only a minor role in European defense.

10 They initially called for a more prominent role for the Western European Union, but that idea was short lived.

11 Croatia is set to become the twenty-eighth member on July 1, 2013.

12 See Nuttall (1992) for a history of EPC. He offers a more optimistic take on the success of EPC than we do.

13 Title V, Article J.1 of the Maastricht Treaty.

14 This unit would soon be split into a Policy Unit and the Joint Situation Centre (SitCen), which opened January 1, 2003.

15 The Lisbon Treaty (Art. 43) expands the list of Petersberg tasks to include disarmament, military cooperation, conflict prevention, and the fight against terrorism.

16 Declaration (No. 30) on Western European Union, B WEU's relations with the Atlantic Alliance (Art. 4). The tasks and institutions of the WEU were subsumed by CFSP in 1999, and the Lisbon Treaty's mutual defense clause now reflects (but does not perfectly mirror) the WEU clause. See TEU Art. 42.7. As a result, the WEU was terminated on July 1, 2011.

17 It is interesting to note how diplomats sought an acceptable military role: "Apart from contributing to the common defense in accordance with the Washington Treaty and Article V of the modified Brussels Treaty respectively, military units of WEU member states, acting under the authority of WEU, could be employed for humanitarian and rescue tasks, peacekeeping tasks, tasks of combat in crisis management, including

peacemaking." Few would have dreamed that this would come to define CFSP (the Swedes reportedly did).

18 Presidency Report on Strengthening of the Common European Policy on Security and Defense. Annex III of the Presidency Conclusions, Cologne European Council, June 3 and 4, 1999.

19 In December 2000 the Nice European Council laid the foundations for the EU and NATO to begin cooperating on crisis management operations across the full range of "Petersberg tasks." Under framework arrangements concluded in March 2003 ("Berlin plus"), the EU is now able to draw on NATO assets and capabilities for this purpose.

20 See Bailes (2005) for an institutional history and Kirchner (2007) for an analysis of the ESS.

21 The ESS implementation report of 2008 did little more than add a set of new threats (cyber and energy security, climate change, and piracy). Report on the Implementation of the European Security Strategy – Providing Security in a Changing World, Brussels: December 11, 2008, S407/08.

22 Kirchner and Sperling (2008) have observed that the willingness of EU member states to cooperate is growing in policies aimed at stabilizing the EU's neighborhood or failing states globally, which is also expressed by the dominance of operations in these areas, but not with regard to responses to threats to the territorial integrity or core member state interests.

23 The specific numbers were calculated in the light of the experiences in Kosovo 1999 and on the basis of estimated requirements for a European force that could operate independently from the USA. For more information, see Lindborg (2001).

24 See Lindström (2007) for a detailed discussion of the EU Battle Groups.

25 The number of troops deployed was 2200 (1785 French and around 200 Swedes). Fifteen other countries, EU members as well as Brazil, Canada, and South Africa, provided strategic airlift, engineers, and helicopters. There were two types of forces in Bunia: a group of special forces (230 troops) and a group of tactical forces. The former consisted of 150 French and 70–80 Swedish troops, a Belgian surgical unit, and French and South African helicopters. A German medical evacuation plane was stationed in Entebbe. The tactical force was made up of lightly armored cars with 90 mm guns. Moreover, 12 French Mirage fighters were stationed at Entebbe, together with a French Atlantique surveillance aircraft, to support the ground forces in the Bunia area (Ulriksen, Gourlay, and Mace 2004: 516–19).

26 Critics charged that the operation only led the militias to regroup further out from town and did very little to address the long-term challenges in

the Congolese province. Indeed, the situation quickly worsened in the fall when the UN took over responsibility for the area (Giegerich 2008: 30). Others criticized the operation for just being a traditional French intervention into African affairs under "European cover" (Homan 2007).

27 Transport and logistics caused problems throughout the operation.

28 In 2006, for instance, the EU Capability Improvement Chart showed that only 7 of 44 shortfalls could be defined as solved (Giegerich 2009). A Capability Development Plan was adopted in 2008 to identify and meet future capability needs, but progress has been slow.

29 Many European countries slashed their defense budgets in response to the financial crisis. Germany reportedly plans to cut its defense spending by a quarter in the period 2012–15. Britain will reduce its budget by 7.5% until 2015. Since the end of the Cold War, defense spending by European NATO members has declined by 20% (Larrabee and Wilson 2012).

30 The difficult aftermath of the NATO intervention in Kosovo (1999), which showed that non-military resources were needed to stabilize this type of crisis situation, reportedly prompted the EU's development of CCM.

31 The definition is Chris Lindborg's, cited in Nowak (2006a: 16). Others have noted that the "actual notion of EU 'civilian crisis management' remains ambiguous and has not been defined in EU documents" (Ioannides 2010: 31; cf. Nowak 2006a).

32 For a concise description of CCM and a history of its development, see Nowak (2006c).

33 For instance, the EU aims for a capacity of 10 000 police officers that can be deployed to faraway theaters (Korski and Gowan 2009: 21). In a separate development, five member states (France, Italy, the Netherlands, Portugal, and Spain) created a European Gendarmerie Force in 2004 (first used in the aftermath of the 2010 Haiti earthquake).

34 "In private, diplomats admit that despite its good intentions, the CHG remains a 'numbers exercise'" (Korski and Gowan 2009: 43). Major and Mölling (2010: 22) call for the establishment of a Civilian Headline Goal 2020.

35 Consensus is not required for (nor are member states involved in) humanitarian aid decisions.

36 For a lucid and detailed overview of ESDP, see Grevi, Helly, and Keohane (2009). See also Hansen (2006: 14–16).

37 Formerly the Group of "RELEX" (Relations Exterieur) Commissioners. Telephone interview with senior official within the Council Secretariat, March 27, 2012.

38 The directorates are "Asia and the Pacific," "Africa," "Europe and Central Asia," "Americas," "Middle East and Southern Neighbourhood," and "Global and Multilateral Issues." These directorates deal with conflict and crisis prevention and coordinate closely with the Commission DG for Enlargement and DG Development.
39 The EU Situation Room is partially built on the former SitCen.
40 "The European External Action Service one year on." Speech by David O'Sullivan, Breakfast Policy Briefing, European Policy Centre, Brussels, Wednesday, January 25, 2012.
41 CSDP is intergovernmental; decisions are taken by unanimity with the right of abstention, but most often are based on a consensus among all the EU states.
42 The CMPD is the result of a merger of directorates VIII and IX of DG External and Politico-Military Affairs (Hynek 2011). The CMPD, CPCC, and EUMS are nominally part of EEAS but maintain a high degree of autonomy (Duke 2010).
43 The International Centre for Information and Intelligence Coordination (former SitCen) continuously monitors and reports on conflicts and crisis situations around the world.
44 The activation of the Operations Centre represents a significant step on the institutionalization ladder. While the option to use its own operational headquarters had existed for years, the EU had relied on the facilities of NATO and member states. By initiating the Operations Centre, the EU has made another incremental step toward independent military capacity.
45 The EDA was established in 2004 to support the Council and the member states in their efforts to improve European defense capabilities in the field of crisis management. It has four tasks: defense capabilities development; armaments cooperation; strengthening the European defense technological and industrial base and defense equipment market; and supporting research. The EDA is governed by a Steering Board of 26 Defense Ministers (Denmark opted out), the EU Military Committee, and the EU Council General Secretariat and is headed by the EU High Representative. In 2009, the EDA had a budget of €31 million and a staff of 109.
46 The member states' updating of the figures of their contributions in a database of national programs and priorities serves to improve the basis for a better division of labor in the provision of national contributions.
47 EDA, "Background Note-Capability Development Plan," July 8, 2008.
48 The EU has a range of long-term (re-)construction instruments, which we do not discuss in this chapter. For more information, see Gourlay (2006a).

49 "Draft Council decision establishing the organization and functioning of the European External Action Service," 8029/10 POLGEN 43, INST 93, Brussels, March 25, 2010.
50 See Cornish and Edwards (2005) and Duke (2008) for a history of EU–NATO relations. For a "typical" pessimistic view, see Valasek (2007). For an alternative perspective, consult Howorth (2009).
51 Key in the early years was the US attitude toward ESDP. The Americans feared ESDP would develop into a defense organization, which would undermine NATO (Smith 2011). The US Secretary of State, Madeleine Albright, formulated the conditions for ESDP in terms of three Ds (no decoupling, discrimination, or duplication). The United States has since grown to support the EU's emerging defense capacity (Keohane 2009: 134). In 2003, an emergency meeting took place between NATO and the EU to discuss the EU's plans. NATO was granted a liaison presence with the EUMS. The ESS subsequently spoke of the "irreplaceable transatlantic relationship" as a "core element" of the international system (Cornish and Edwards 2005: 812).
52 A big problem is the parallel sets of capability and procurement lists, which impedes interoperability between both organizations.
53 The Berlin Plus agreement also made it possible for non-EU NATO members to partake in the EU decision-making process on EU-led operations that make use of NATO resources (Riggio 2003; Haine 2004).
54 In practice the participants of these meetings are 26 EU ambassadors (from all member states except Denmark, which has opted out from CSDP) and 28 NATO ambassadors.
55 Since 2009 NATO has had 28 member states including 21 EU member states, Albania, Canada, USA, Iceland, Norway, and Turkey. The EU has six non-NATO members: Austria, Cyprus, Finland, Ireland, Malta, and Sweden. Austria, Finland, Ireland, Malta, and Sweden are members in NATO's Partnership for Peace (PfP). Only one EU member state (Cyprus) does not have formal links with NATO.
56 Council of the European Union (2007), "The Africa–EU Strategic Partnership – A Joint Africa–EU Strategy," 16344/07 (Presse 291). Lisbon, December 9, 2007.
57 €92 million has been used to support AMIS. Council of the European Union (2006), "European Union concept for strengthening African capabilities for the prevention, management, and resolution of conflicts," 11316/06. Brussels: July 7, 2006. "Council Conclusions on a Policy framework for Security sector reform" (2736th General Affairs Council Meeting, June 12, 2006); European Commission (2007), "Press Release. Commission of the European Communities, Commission/Council Secretariat Joint Papers, Beyond Lisbon. Making the EU–Africa Strategic

Partnership work." COM (2007) 357 final. SEC (2007) 856. Brussels: June 27, 2007.

58 For instance, the High Representative cooperated closely with UN Secretary-General Kofi Annan in the process that led up to the UN's request for EU help in Congo (Hendrickson, Strand, and Raney 2007: 7).

59 In 2010, the European Commission restructured its bureaucratic organization, and in the process combined humanitarian aid policymaking with that of civil protection into a new DG ECHO. The operational capacities of each unit (including their respective crisis rooms and capacity inventories) were also merged.

60 The EU spent €5.7 billion on humanitarian aid and civil protection between 2007 and 2013 – most of it on humanitarian assistance.

61 See Brusset and Tiberghien (2002) for an informative chapter on ECHO's institutional history.

62 The main difference between the Civil Protection Mechanism and ECHO is that the aim of the former is to provide for assistance in the intervention phase: "in the first week, or in maximum the first fourteen days." The Civil Protection Mechanism foresees that EU experts can be sent out within 12 hours after an event. Whereas ECHO finances the rebuilding of damaged infrastructures, EU civil protection experts offer "first aid" after an emergency. Civil protection assistance is only provided in response to a request from the relevant state. Humanitarian aid through ECHO funds NGOs to provide emergency assistance regardless of any request or agreement from the country hit by disaster.

63 The ERC merges the Monitoring and Information Centre (MIC) and the DG ECHO's crisis room. The merger is scheduled for 2013. The response capacity will be based on pre-committed member states' assets and pre-agreed contingency plans.

64 This is clearly described in a document developed by the Crisis Management and Planning Directorate (CMPD) and sent to member states' delegations a few days before the Foreign Affairs Council on January 25. See the Note from the Council Secretariat (2010): "Proposals for coordination of EU contributions in response to specific UN requests for assistance to the situation in Haiti." 5610/10. Brussels: January 21, 2010.

65 Joint Motion for a Resolution on the recent earthquake in Haiti, presented on February 9, 2010.

66 For instance, SitCen placed a small team of liaison officers in the UN Stabilization Mission MINUSTAH during its response to the Haiti disaster. In DRC, the French command stationed a civil–military liaison officer with the humanitarian assistance agencies (Homan 2007).

67 See the House of Lords Report (2009a) for a blistering critique of the lack of cooperation.
68 For the challenges of a military mission in Africa, see Seibert (2007).
69 Council of the EU, Secretariat to the political and security committee, "Increasing the flexibility and usability of the EU battle groups," 14036/1/09, REV 1, Brussels, October 29, 2009.
70 A formula known as the "Athena mechanism," by which member states contribute on the basis of their respective GDPs, covers some common costs (about 10% of a mission's overall costs).
71 The CFSP budget only counts for 3.5% of the budgetary allocations for EU external relations at large: €243 million is allocated to CFSP, €2.4 billion to development, and €6.1 billion for European Neighborhood and Partnership Instrument (Grevi and Keohane 2009: 90).
72 When a delegation of MEPs visited the military operation in the DRC in 2006, they expressed amazement at the sheer length of the command chain for such a small operation (Simon and Mattelaer 2011: 22).
73 For a picture of this complex process, see Korski and Gowan (2009: 77).
74 The Commission took the Council to Court (the European Court of Justice: Case Number C91/057) over a Council decision (2004/833/CFSP of December 2, 2004) regarding the funding of a program to limit the proliferation of small arms and light weapons (SALW) in West Africa. In the view of the Commission, supported by the European Parliament, this was a matter to be funded under the EDF (European Development Fund) within the framework of the Cotonou Agreement and the Council should not have done this within pillar two. The Council, supported by France, Spain, Sweden, the UK, and the Netherlands, argued in a totally opposite direction, denying that SALW could be a development issue at all and that it is only a security issue.
75 See Bossong (2012) for a detailed analysis of how the EU has learned, and should learn, from its civilian crisis missions.
76 An open question is whether the preferred remedy – address coordination failures by adding coordinative institutions – has made the problem worse (Khol 2006).
77 "Towards a stronger European disaster response: the role of civil protection and humanitarian assistance (Text with EEA relevance)." COM (2010) 600 final. Brussels: October 26, 2010.
78 See Hynek (2011) for a detailed description of concepts (CIMIC and CMCO) that are aimed to further collaboration between civil and military crisis management.
79 Communication from the Commission to the Council, the European Parliament, the European Economic and Social Committee and the Committee of the Regions (2005). "EU Disaster and Crisis Response in Third

Countries." COM (2005) 0153 final. Brussels: April 20, 2005, pp. 5–7; Communication from the Commission to the Council, the European Parliament, the European Economic and Social Committee and the Committee of the Regions (2005). "Improving the Community Civil Protection Mechanism". COM (2005) 0137. Brussels: April 20, 2005, pp. 7–8; European Commission (2005). "EU Civil protection assistance in South East Asia." Memorandum, MEMO/05/6. Brussels: January 11, 2005.

80 The European Gendarmerie Force (EGF) is an initiative of five member states (France, Italy, Netherlands, Romania, Spain). The idea behind the EGF cooperation is to allow a fast deployment (within 30 days) of interoperable EU police forces with a military status.

81 See Sally McNamara, "EU Foreign Policymaking Post-Lisbon: Confused and Contrived," Backgrounder #2388, March 16, 2010. Heritage Foundation. May be found at http://report.heritage.org/bg2388.

82 The foreign policymaking powers of the permanent President, as formulated in the Lisbon Treaty, allowed him both to shape the EU's overall strategic objectives in external affairs and to call extraordinary Council summits in the face of international crisis.

83 Mark Rhinard and Arjen Boin, "Relief effort reveals some uncomfortable truths," *The European Voice*, January 28, 2010. See also, Bruno Waterfield, "Lady Ashton under fire over 'EU visibility,'" *The Telegraph*, January 22, 2010. "La diplomate en chef de l'UE, critiquée por sa gestion de crise," *Le Point*, January 19, 2010.

84 This impression was reinforced by her subsequent attempt to defend her decision by stating "I am neither a doctor, nor a firefighter" (as quoted by McNamara, "EU Foreign Policymaking Post-Lisbon: Confused and Contrived"). See also Honor Mahony, "Ashton under fire for not going to Haiti," *EU Observer*, January 19, 2010; Joshua Chaffin, "Lady Ashton needs someone versed in the dark arts of spin," March 29, 2010, ft.com/brusselsblog at http://blogs.ft.com/brusselsblog.

85 "La diplomate en chef de l'UE, critiquée por sa gestion de crise," *Le Point*, January 19, 2010. See also Rhinard and Boin, "Relief effort reveals some uncomfortable truths."

86 Tony Barber, "Poll backs bigger world role for EU," *Financial Times*, September 7, 2007.

87 According to Missiroli (2008: 1), the ESS "arguably ranks as one of the best EU documents ever written" and "has stood the test of time to an unusual extent." Grevi (2009: 24) speaks of a "landmark development."

88 After having initially attempted to play broker, Solana was reportedly sidelined during the months leading up to the Operation Iraqi Freedom in winter 2003 (Dempsey 2003: 47; Allen and Smith 2004: 95).

89 The member states closely monitor the EU's security institutions. While policymakers and administrators in the field may have substantial autonomy, scrutiny at the political level is intense. The Political and Security Committee carefully ran the first missions to keep all member states on board. Having diplomats lead missions may not have been very efficient from a technical perspective, but it was sound from a political point of view.

90 One might say that the EU has been a trainer of policemen rather than a policeman.

91 This section is largely based on Grevi, Helly, and Keohane (2009). See the CSDP home page for a most recent overview: http://www.consilium.europa.eu/eeas/security-defence/eu-operations.

4 *Managing transboundary crises*

The gradual emergence of EU capacity

The countries of Western Europe do face rather similar threats and quite a few of them have to be faced collectively.

(Lord Hannay of Chiswick, House of Lords 2011b: 218)

I do not think that the Union has decided how it wants to manage its response to crises.

(William Shapcott, House of Lords 2011b: 16)

Introduction: the prospect of transboundary crises

The previous chapters described how the EU has been building capacity to coordinate a joint response to overwhelmed member states and other disaster-stricken areas; moreover, they demonstrated a growing capacity to send civil–military missions to worldwide hot spots. This chapter focuses on a different type of crisis, one that affects multiple member states and, when left unaddressed, can threaten the political fabric of the European Union. This chapter documents the EU's efforts to deal with transboundary crises (Boin and Rhinard 2008).

These crises find their origin in the very essence of what the EU seeks to achieve. Since 1957, the European Union has prioritized the "four freedoms": the free movement of goods, services, capital, and labor across borders (Art. 12 EU Treaty). Physical and technical barriers were lowered, regulations were harmonized or made compatible, and critical infrastructures were tied together. In the 1980s, member states agreed to create a single economic market in Europe. The integration of Europe has made EU member states increasingly vulnerable to what could once be considered "foreign" or "local" problems in distant places (OECD 2003, 2011; Missiroli 2005; Sundelius 2005).

Many of the systems that sustain basic societal functions (e.g. energy grids, transport networks, food distribution, financial flow structures) now reach across European borders. As a result, an incident in one corner of Europe can turn into a crisis for the entire continent: a nuclear explosion, an epidemic, a power outage, a wave of immigrants, a blocked oil pipe, a cyber attack, an ash cloud, or a budget problem of a member state – these are all crises that have demonstrated potential to reach far beyond geographical and functional boundaries.

Transboundary crises unfold across borders and have widespread consequences (OECD 2011). Their cascading nature and the insidious knock-on effects typically outstrip local coping capacity and resist unilateral responses. Transboundary crises are unusually complex and very hard to manage. They cannot be managed by member states alone (Ansell, Boin, and Keller 2010). If there is one type of crisis deserving of EU capacities, we argue, it is the transboundary type.

This chapter identifies and describes a variety of transboundary response capacities at the EU level. While quite a few capacities relevant to transboundary crises have emerged at this level, they are not always easy to identify. These capacities are not concentrated in one agency or policy directorate. In contrast to the other crisis types studied in this book, most transboundary crises have no specified policy "home" inside the EU institutions. They do not even have a common label (few people speak of transboundary crisis management or anything resembling the term).

We searched across the Union for capacities – policies, methods, organizations – that can be used to facilitate a joint response to a common and urgent threat. In this chapter, we present our findings and explore how these capacities emerged. We begin by focusing on those policy sectors that we suspect have developed transboundary crisis management capacities (even if these capacities are not labeled as such). In addition to "vertical" (sector-specific) capacities, we look for "horizontal" (sector-spanning) measures that can be used to manage transboundary threats, crises, and disasters. We conclude this chapter with a reflection on the potential for further institutional development of these capacities.

Building transboundary crisis management capacity: a brief policy history

Before the EU began building capacity to assist overwhelmed member states (Chapter 2) and before it began sending civil and military missions to hot spots around the world (Chapter 3), the member states had already discovered the deep challenges posed by what we refer to as transboundary crises. They also learned what happens in the absence of adequate response capacities. Let's take a brief look at two rather vicious "common threats" (as they are sometimes referred to in EU parlance) that have bedeviled member states and EU institutions.[1]

Radioactive clouds over Europe

In the early morning of April 26, 1986, reactor 4 of the Chernobyl nuclear power plant in Ukraine suffered a massive nuclear chain reaction. An explosion tore off the top of the reactor building exposing the reactor core, which allowed the dispersion of large amounts of radioactive particulate and gaseous debris (Medvedev 1992).

The explosion released at least 100 times more radiation than the atomic bombs on Hiroshima and Nagasaki. Nearby regions in Belarus, Ukraine, and Russia received much of the fallout, but in nearly all countries in the northern hemisphere traces of radioactive deposits were found (Liberatore 1999). People feared food might be contaminated and countries began to close their borders to produce from "suspect" countries.

The Commission was slow to respond and member states made little effort to coordinate. The EU's response was shaped by different levels of competence given to the EU by member states in the two policy domains most affected: health and trade. The protection of health was not an EU competence in 1986. Trade, however, was an "exclusive" EU competence and a high priority for the EU (Marples 1986). As soon as Chernobyl affected the free circulation of goods within the EU and trade with external countries, the Commission could legally intervene to standardize measures and issue basic guidelines. But the Commission was slow to respond even to the trade dimension of Chernobyl (Liberatore 1999: 205).

One reason for this was the lack of adequate communication channels for use during a nuclear crisis. Communication lines for nuclear issues did exist, via a network of national authorities founded by the Euratom Treaty. Yet that network was not intended for emergency situations (Liberatore 1999: 205). Moreover, cooperation on nuclear crisis management was conducted via the International Atomic Energy Association (IAEA).[2] As a result, the Commission had precious little information regarding the extent of the fallout and the implications for food and agriculture in Europe.

The disaster revealed gaps in coordination and information exchange amongst European states and served as an impetus for new legislation in areas ranging from health to trade and from emergency response to research. New powers were given to the Joint Research Centre in Ispra, Italy, to establish the "Radioactivity Environmental Monitoring" database to collect and organize readings of contamination within Europe.

Mad cows in the UK

In the early 1990s, a rare but horrible neurological affliction, Creutzfeldt–Jakob disease (CJD), became linked to a disease in cows (Bovine Spongiform Encephalopathy [BSE] or "mad cow disease"). The UK acknowledged the first cases of BSE in 1986. On March 20, 1996, the British government acknowledged a possible connection between BSE and CJD. Because there was an incubation period of several years, anybody who had ever eaten beef from the UK could be infected – even if the chance was infinitely small (Grönvall 2001; Millstone and Van Zwanenberg 2001).

Almost immediately after the British announcement, several EU countries closed their borders to British beef. The "mad cow" scare, fueled by frenzied media reporting, spiraled out of control and a full-fledged panic took hold of European consumers.

Initially defined as a veterinary issue, and in the light of the disease's implications for the internal market, the Commission's DG Agriculture and Rural Development took the lead (Grönvall 2001). The BSE problem clearly had implications for public health but the Commission's DG Health and Consumer Protection was not established until 1999 and, thus, was not involved in the EU's early crisis decision making.

DG Agriculture made use of the standard operating procedure for a veterinary problem: it convened the Commission's Scientific Veterinary Committee (ScVC) to advise what action should be taken. The ScVC concluded that the United Kingdom had taken appropriate measures and no ban of UK beef would be necessary. In addition to the ScVC, which contained independent experts, the Council's Standing Veterinary Committee (SVC), consisting of national representatives, became involved. The SVC voted 14 to 1 (the UK casting the dissenting vote) in favor of a ban on UK beef and related products (gelatin, tallow, and bull semen).

The bungled EU response plunged the EU into one of the deepest crises it had ever experienced. The UK, under domestic pressure to retaliate against the imposed ban, announced a so-called "non-cooperation policy" (Grönvall 2000; 2001). The UK refused to cooperate with other member states in the European institutions, particularly by blocking measures in the Council. The UK government launched proceedings against the European Commission in the European Court of Justice, arguing that the ban was unjustified and violated due process. In July 1996, the Court ruled against the UK government. By that time both national and European officials deemed the contamination risk to have sufficiently subsided and the EU ban was partially lifted (Grönvall 2000: 77–8).

The EU's response to the BSE crisis has been widely described as a failure (see Phillips Inquiry: Phillips 2000; Vincent 2004). Following the BSE crisis, the Commission took some major steps toward dealing more effectively with this type of transboundary crisis (Ansell and Vogel 2006; Matzén 2008; Macmaolain 2011). Policy portfolios were reshuffled, DG Agriculture was restructured, and the newly established DG for Health and Consumer Protection was invested with new crisis management competences. These changes resulted in an improved response to the dioxin scandal that broke out in 1999 (Olsson 2005; Staelgrave and 't Hart 2008).

Further steps were taken to improve crisis management and to clarify the boundaries between risk assessment and risk management across the EU institutions. In 2002, the European Food Safety Agency (EFSA) was created for risk assessment and risk communication on food safety questions. The agency operates the Rapid Alert System for Food and Feed (RASFF), which comprises a network of monitors in every member state and flags dangerous trends in the food supply.[3] EFSA has no risk management responsibilities; such decisions remain in the domain of the Commission and member states.

The EU's fragmented approach to managing transboundary crises

These early encounters with transboundary crises exposed the EU's inability to produce a quick, coordinated, and joint response. These crises also demonstrated what the EU would need, at a minimum, to deal with this type of unfolding threat: a capacity to share information, to foster speedy decision making, and to speak with one voice (see also Ansell, Boin, and Keller 2010).

As we will see, the EU has made great strides in the wake of subsequent transboundary crises. These steps were mostly sector based. The EU has been less successful in creating generic, cross-sectoral crisis management capacity. This explains why the EU continues to be surprised and confounded by new transboundary crises such as the Icelandic ash crisis, the EHEC outbreak, and the raging financial crisis.

It is, of course, much harder to develop generic capacities than it is to implement specific measures to prevent a certain crisis from ever happening again. It is this tension between the need for generic crisis management capacity and the multitude of threat-specific measures that characterizes this emerging policy field. The EU is still wrestling to provide an encompassing mission for these capacities and an administrative apparatus to coordinate its various organizational and policy tools.

Many of the efforts to build both threat-specific and generic capacities find their origin in the EU's policy area of Justice and Home Affairs (JHA) (dubbed the "Third Pillar" of EU decision making, until the Lisbon Treaty abolished the pillars) (Walker 2004; Kaunert 2010a; Monar 2010).[4] Confronted by the unintended consequences of their efforts to create open borders, the member states were forced to contemplate "flanking measures," protective steps taken within the margins of existing single market policies, to maintain safety and security in an increasingly borderless Europe. As the EU ventured out beyond flanking measures to tackle internal security directly, these issues verged upon the borders of national sovereignty. Internal security is still a national responsibility enshrined in Article 72 of the TFEU; efforts to enhance transboundary coordination therefore have moved slowly.

Justice and interior ministers of EU member states began cautiously working to enhance collaboration on matters related to internal security in the early 1970s. The member states have since created a "complex and often murky decision-making structure" at the EU level to manage security- and immigration-related challenges. Through this structure, national officials "grind out an ever-growing range of legal and practical 'flanking measures' needed to deepen cooperation between police, judges, border guards, customs officers, immigration officials, and other public authorities. Taken together, these decisions are known as EU 'Justice and Home Affairs' policy" (Brady 2011).

This JHA area has seen a widening (but never formally agreed on) definition of internal security. The initial initiatives were concerned with enhancing collaboration on criminal justice and border matters. After the 9/11 events, counter-terrorism and the protection of critical infrastructures gained prominence. The idea of internal security eventually widened to include a wide variety of threats to the safety and well-being of citizens (see also Chapter 3). Policy initiatives began to include the management of common threats and crises that reach across boundaries.

A casual reading of Third Pillar action programs illustrates this shift. The first five-year program (Tampere), launched in 1999, formulated a set of initiatives to further collaboration on crime fighting and developing a common asylum and immigration policy. Its successor (the Hague Programme) was formulated in the post-9/11 world. It identified the protection of critical infrastructures as a priority, along with "consequence management" of terrorist attacks. It also gave rise to two generic crisis management capacities: Argus and the Crisis Coordination Arrangements (CCA).

Argus

Argus is a web-based system that facilitates the sharing of information across directorates in the Commission (Kjellen 2009).[5] Argus not only links directorates and units together, it also serves as a "network of networks" bringing together the many rapid-alert systems operating independently in various Commission directorates. The Commission's Security Office is the operative point of contact for Argus, with staff on call around the clock. The policy point of contact for Argus is a unit within the Secretariat-General of the Commission, established in 2008 as part of the Secretary-General's personal drive to improve coordination in crisis management matters.

Warnings are categorized according to crisis levels I or II (Kjellen 2009: 79). Level I warnings are sector specific and can be managed by the directorate most immediately involved. Other directorates are simply informed. Level II warnings affect multiple directorates and, thus, require operational coordination. Only the Commission President, or his designee, has the authority to authorize a level II crisis warning.

A level II warning will activate the Commission's Crisis Coordination Committee, which is comprised of high-level Commission officers (normally directors-general) with responsibility for the affected sectors. The Crisis Coordination Committee will take crisis decisions on behalf of the Commission under the direction of the Commission President.

The Argus system sparked an initiative to hold regular meetings among crisis-related administrators in the European Commission, organized by the Commission Secretariat-General, to discuss standard operating procedures for triggering Argus, for sharing information and knowledge on sectoral threats, and for identifying gaps and overlaps in capacities.

Crisis Coordination Arrangements

As part of the Hague Programme, the European Council invited the Council and the Commission to "set up within their existing structures... integrated and coordinated EU crisis-management arrangements for crises with cross-border effects within the EU."[6] The 9/11 attacks had contributed to a growing awareness that, outside of the normal legislative process, there was no procedure for national political representatives to join with the EU's institutional leadership to ensure a coordinated set of responses to such major events. This declaration, contained in Section 2.4 of the Hague Programme, is one of the few explicit acknowledgments in EU documents regarding the importance of creating decision structures for transboundary crises.[7]

The Crisis Coordination Arrangements (CCA) provide for a way to deal with the strategic and political dimensions of crisis decision making. A 2007 CCA manual sets out procedures for intergovernmental and inter-institutional coordination to facilitate common public communication and crisis management action at the EU level.[8] But the CCA are still in a formative stage and have not been used very often. They were put on "standby" during the wars in Lebanon (July 2006) and Georgia (August 2008), were placed on "alert mode" during the Mumbai terror attacks (November 2008), and were almost triggered during the recent financial crisis. The CCA are exercised annually.[9]

Toward consolidation: the Stockholm Programme and the Internal Security Strategy

The third action program, known as the Stockholm Programme, set out a range of initiatives for the period 2010–14. The tone of the document shifted from an emphasis on criminal justice and immigration to building a "Europe that protects." The 84-page document calls for a comprehensive, *use all means available* approach to protect EU citizens from a wide range of threats: "We need to address all the common security threats from terrorism and organized crime, to safety concerns related to man-made and natural disasters."[10] To impose some order on the wide variety of initiatives found in the Tampere, the Hague, and Stockholm Programmes, the latter program called for a comprehensive Internal Security Strategy (ISS).

In March 2010, the Council adopted the ISS. In contrast to the externally focused European Security Strategy discussed in Chapter 3, this document is not very compelling.[11] It offers a rather vague summation of various threat types without a clear pathway toward enhanced protection of EU citizens.[12] A few months later, however, the Commission improved on the Council document by producing a follow-up document (*The Internal Security Strategy in Action*).[13] Drafted largely by DG Home, this document identified five core threats on which the EU should focus:[14]

(1) Disrupt international crime networks.
(2) Prevent terrorism and address radicalization and recruitment.
(3) Raise levels of security for citizens and businesses in cyberspace.
(4) Strengthen security through border management.
(5) Increase Europe's resilience to crises and disasters.

The first four threats had consistently emerged in JHA documents and feature prominently in previous action programs. The addition of the fifth action directive – increasing resilience – highlights the expanding conceptualization of internal security. It also underlines the ad hoc nature of formulating strategies in this domain.

An accompanying Commission document – outlining the financial support for internal security initiatives – illustrates the conceptual ambiguity surrounding the stated ambition to enhance resilience, crisis, and disaster management.[15] The document sets out by referring to "a coherent and comprehensive approach to law enforcement cooperation, including the management of the Union's borders," which suggests no role at all for crisis management or resilience. Indeed, it explicitly refers to Article 3(2) of the Lisbon Treaty, which states that "the Union shall offer its citizens an area of freedom, security, and justice without internal frontiers, in which the free movement of persons is ensured in conjunction with appropriate measures with respect to external border controls, asylum, immigration, and the prevention and combating of crime." All this suggests that crisis management and resilience are "add-ons."

Even if they are not, it remains unclear what the Commission, more precisely DG Home, envisions when it refers to crisis management and resilience. According to a definition offered in Article 2 (f), "risk and crisis management means any measure relating to the assessment, prevention, preparedness, and consequence management of terrorism and other security-related risks." Article 2 (j) refers to an "emergency situation," which means "any security-related incident or newly emerging threat which has or may have a significant adverse impact on the security of people in one or more Member States." The document further announces that the EU will fund "secure links and effective coordination between sector-specific early warning and crisis cooperation actors at Union and national level... *including those for health, civil protection, and terrorism*" (Annex p. 24). But a few pages down (p. 31), it introduces a division between natural disasters and unintentional man-made disasters (which are for civil protection) and intentional, man-made disasters (reserved for Internal Security).

What we see here, in just one document, is a multitude of seemingly related concepts (risk, emergency, crisis, terrorism) that are not consistently used or logically related. A core concept such as resilience simply remains undefined (and is hardly mentioned).

The Commission document *The Internal Security Strategy in Action* is much more coherent as it sets out several actions that promise to enhance the EU's capacity to manage transboundary crises. In addition to initiatives to enhance cooperation in the area of criminal justice (including counter-terrorism and cybercrime), it offers sensible proposals to develop the EU regime for aviation and maritime security, common risk analyses, a coherent risk management policy, and a "link-up" of different situation awareness centers.

The EU has made funding available for these initiatives. For the period 2014–20, the Commission is proposing an overall Home Affairs budget amounting to €10.9 billion, which represents an increase of almost 40% compared to the total budget for the period 2007–13. This amount will be divided into two funds: a new Asylum and Migration

Fund with an overall budget of €3869 million and a new Internal Security Fund worth €4648 million.[16]

The implementation of ISS on the Commission side is in the hands of DG Home, a newly organized DG still finding its footing. The Lisbon Treaty also created a Council committee, COSI, which should help to coordinate the implementation of ISS. COSI is a high-level working group: officials attend from capitals, meet for a day, and decide on various courses of action that are proposed to the JHA Council. While conceived in some quarters to be the internal security equivalent of the PSC (discussed in Chapter 3), COSI has reportedly experienced a slow start and found it hard to develop a clear identity and a convincing modus operandi. Moreover, it operates in a dense field of "competing" bodies (see House of Lords 2011a: 51–2).

In summary, we can say that the field of internal security is a work in progress. The EU has developed a plethora of organizations, policies, and instruments that may enhance a joint response to a transboundary threat.[17] The challenge, taken up in the *Internal Security Strategy in Action* document, is to impose some order on all this variety.

To get a better idea of the variety, and the challenges it poses for any integrative effort, we will now take a look at the most prominent policy sectors in which transboundary crisis management capacities are being developed: counterterrorism, border management, communicable diseases, critical infrastructures, and finance.

Countering terrorism

European-wide cooperation on counter-terrorism commenced in the terrorism-plagued 1970s, within the framework of the Trevi Group (Von Hippel 2005; Bures 2011).[18] The Trevi Group (established in 1976) consisted of high-ranking officials from interior and home ministries, along with the heads of European security and intelligence agencies. Their informal gatherings facilitated the exchange of information and mutual assistance on terrorism and international crime. With the adoption of the Maastricht Treaty in 1992, the group was formally incorporated into the EU's third pillar.

At the Madrid European Council session of 1995, heads of state signaled their converging understanding that terrorism should be treated as a common threat to democracy and economic and social development. Further demonstrating the rise of terrorism on the EU's agenda and list of competences, the Treaty of Amsterdam included a specific reference to the fight against terrorism, stating that "preventing and combating crime, organized or otherwise, in particular terrorism," is one of the tasks of the Union (Art. 29 para. 2).

After the September 11, 2001, attacks, the European Council pledged to increase counter-terrorism cooperation (Dalgaard-Nielsen and Hamilton 2006; Zimmermann 2006; Argomaniz 2011). The EU adopted a "Framework Decision on Combating Terrorism," which included a common definition of terrorism.[19] The EU also adopted an "Action Plan to Fight Terrorism" and a program aimed at chemical, biological, radiological, and nuclear (CBRN) threats (Bossong 2008).[20]

In 2003 the Council adopted the European Security Strategy (discussed in Chapter 3), which asserted that "terrorism poses a growing strategic threat to the whole of Europe" and that "Europe is both a target and a base for terrorism." The Hague Programme (2004) emphasized enhanced prevention, preparedness, and response to terrorism offences, reduction of terrorists' financial and economic resources, and exchange of investigation information and protection of critical infrastructure programs.

The 2004 Madrid bombings underlined the importance of joint action and helped to shape a perception that an attack on one member state is an attack on the Union. In the words of Commission President Barroso:

> An attack against an EU citizen is an attack against all Member State citizens. An attack against a Member State is an attack against the Union. It is an attack against the values we stand for. Terrorism seeks to destabilise societies by creating tension, fear and panic. Reactions to the Madrid events of last year made it clear that a terrorist attack on any part of the EU affects the EU in its entirety. This is natural because the EU is a Union of nations whose governments have signed a Constitutional Treaty that solemnly announces the principle of solidarity. The security requirements of such a Union reach beyond borders. [21]

On March 25, 2004, two weeks after the Madrid bombings, the Council called for a revised Plan of Action.[22] The revised Plan of Action recommended boosting the authority of Europol (see below) and authorized a specialist anti-terrorist unit to improve the sharing of police data and information and to conclude operational agreements with countries such as the United States. It also proposed further measures aimed at the suppression of terrorist financing, including several instruments to allow for the freezing of assets and the tracking of monetary transfers across borders.[23]

The Plan of Action included plans for the implementation of the European Arrest Warrant (EAW), which had first been introduced in a Council Framework Decision of June 13, 2002.[24] The EAW requires EU member states to recognize and respond to surrender requests from fellow states. It is an instrument of "mutual recognition," a principle by which member states must accept and recognize key measures in each other's judicial systems, and was envisioned to serve as the backbone for further JHA cooperation (Dittrich 2005).[25]

That same year, the European Council appointed Mr. Gijs de Vries to the new position of Counter Terrorism Coordinator. His responsibilities included the coordination of Council efforts to combat terrorism, as well as updating all the Union's counter-terrorist instruments.

Other actions, which can in part be seen as follow-up to the Madrid bombings, were the Council's updating of the EU list of terrorist organizations and individuals connected to terrorist activities.[26] In addition, member states agreed on a comprehensive approach to regulating aviation security standards at all EU airports.

On the morning of July 7, 2005, three bombs targeting the London transport system exploded within a minute of one another. An hour later, a fourth explosion ripped open a double-decker bus at Tavistock Square. The attacks killed 52 people and wounded more than 770.

The London bombings led to a number of initiatives (many already in the legislative pipeline), including a framework decision on simplifying exchange of information and intelligence between law enforcement authorities, a draft framework decision on the retention of telecommunications data, and a draft framework decision on the European Evidence Warrant.[27]

Institutional structures of EU counter-terrorism

Since the 1970s, the EU has also built organizational capacity to coordinate the interactions between security agencies of the member states (Fägersten 2010). The creation of DG Justice, Freedom and Security (DG JLS) in 1999 marked an important step on the institutionalization ladder. The evolution of two agencies, Europol and Eurojust, reflect further growth of institutionalization. Both agencies started their organizational lives as "toothless tigers," but they have gradually developed into respected organizations in and through which specialists share information and initiate cooperation among relevant member state agencies (House of Lords 2011a).

Europol, the European Union's law enforcement cooperation organization, was created as part of the 1992 Maastricht Treaty. The original remit of Europol was extremely restricted, due to disagreement among member states as to its exact purpose. The organization focused initially on drug interdiction cooperation, but Europol's mandate has expanded progressively since the 1990s (Lavenex and Wallace 2005; Fijnaut 2006; Kaunert 2010b). By 2002, the mandate of Europol was extended to deal with all serious forms of international crime.

Since 2001, terrorism has assumed special significance for Europol, with several high-level working groups in the Council now incorporated in the organization's headquarters in The Hague. A counter-terrorism unit analyzes information on terrorism developments and produces threat and risk assessments.

The establishment of Eurojust, the European Judicial Cooperation Unit, in 2002, signaled the intentions of member states to enhance cooperation in investigating and prosecuting serious cross-border crime. While its formal authority is still limited, this unit has become a hub for information exchange among European prosecution agencies (Thwaites 2006; Van den Wyngaert 2006).

A substantial number of working groups and preparatory committees can be found in the area of counter-terrorism cooperation (Dittrich 2005). One more peripheral venue in which counter-terrorist measures are discussed is the Aviation Security Regulatory Committee, initiated in 2002. The Joint Situation Centre (now placed in the EEAS (see Chapter 3) assists EU member states with analysis of international threats such as terrorism.[28] The EU Police Chiefs' Task Force (established in 2000) provides a forum in which senior police authorities meet regularly to coordinate operational approaches to transnational crime problems. The COSI is the most recent addition to this already dense field.

The EU Health and Security Committee has established a 24-hour, 7-day-a-week Rapid Alert System for Biological and Chemical Attacks and Threats (RAS-BICHAT) (Sundelius and Grönvall 2004). This alert system relies heavily on the member states' surveillance systems for the occurrence of infectious diseases. It is operated by the European Centre for Disease Prevention and Control (see page 118 ff). Finally, inventories of laboratories and their diagnostic capabilities have been made to facilitate reliable and timely detection of likely biological agents.

Managing borders

The management of external borders has a long EU history.[29] The story begins in the early 1970s, when the 1973 oil crisis and subsequent economic depression led to widespread migration. With anti-immigrant sentiment rising, most European countries put a stop to the legal admission of foreign workers after years of permissive consent (Pastore 2007). Immigrant flows predictably shifted toward the more "permissive" countries (Alink 2006). Although such events did not prompt common policies in the 1970s, they did nurture efforts to coordinate more closely and regularly, albeit outside of the EU framework (Pastore 2004: 94).

The Single European Act in 1986 turned the EU into "an area without internal frontiers in which the free movement of goods, persons, services, and capital is ensured" (SEA, Article A). The prospect of open borders quickly drew attention to the problem of differentiated immigration and asylum procedures within Europe and highlighted the need not only to improve coordination but also to formulate common policies (Barutciski 1994; Monar 2010).

Policy cooperation deepened, but in an intergovernmental fashion. The first Schengen Agreement (1985), which took a step toward abolishing border controls for individuals moving between signature countries, included only five of the then-seven member states. The second Schengen Agreement (1990) took several steps further, establishing plans to completely abolish internal border controls, to create common external border policies, and to adopt common visa policies for third-country individuals. Reacting to the problem of "asylum shopping," Schengen countries signed the Dublin Convention (1990) by which the first country receiving an asylum application must examine and deal with the claim.

The Maastricht Treaty (1992) brought immigration policy firmly within the EU framework, although the policy process retained an intergovernmental mode of decision making as part of the third pillar (Balch and Geddes 2011).

Yugoslavia slid into civil war in June 1991. By the end of 1993, more than 500 000 refugees had fled abroad (Carmichael 2002). As migrants began pouring over the borders into Western Europe, some countries adopted special measures to accommodate them (Barutciski 1994: 32). Other member states refused to lift normal asylum procedures or offer temporary protection. The effect was to drive migrants to European states which still offered refuge (Mazower 2002).

The EU was widely criticized for its understated role in the Balkans, initially in the attempt to dampen the conflict (see Chapter 3) and again as the refugee crisis unfolded (Crawley 1993). Fragmented legal arrangements, poor coordination of tightening and loosening of restrictions, and a strategy of sending aid and resources to the Balkans rather than manage migration flows within Western Europe sparked objections from human rights groups and the UNHCR.

With the adoption of the Amsterdam Treaty (1997) EU immigration policy became fully "communitarized" (Balch and Geddes 2011). Decision making on immigration policy was moved from the third to the first pillar, thus subjecting it to the Community Method (e.g. active Commission participation and decision making by majority vote).[30] At the turn of the millennium, the Tampere European Council (1999) adopted a common migration and asylum policy for the first time in the EU:

> The European Union needs a comprehensive approach to migration addressing political, human rights, and development issues in countries and regions of origin and transit. This requires combating poverty, improving living conditions and job opportunities, preventing conflicts and consolidating democratic states and ensuring respect for human rights, in particular rights of minorities, women and children. To that end, the Union as well as Member States are invited to contribute, within their respective competences in the Treaties, to a greater coherence of internal and external policies of the Union.[31]

In 2002, at the Seville summit on combating illegal immigration, an action plan was adopted with a variety of measures to guide the development of an integrated EU approach to border management (Boutruche 2003).[32] Still in effect today, it provides for a set of operational measures by member states aimed at improving coordination and cooperation, achieving a common integrated risk analysis, securing personnel and hardware for border management, and working toward common legislation on burden sharing (Illies 2009).

Most of these measures fall into two categories. The first category includes standards that all external border efforts must follow.[33] The second category involves measures for operational coordination and cooperation. These measures were consolidated largely through the creation of Frontex (see below).

In addition to coordinating policies of member states, the EU adopted a strategy of "externalizing" the problem (Uçarer 2002; Lavenex 2004). In 2004, the Commission developed the notion of "Regional Protection Programmes" to support areas of the world with high levels of refugees (Haddad 2008).[34] It proposed to establish asylum-processing centers in the "buffer" countries surrounding the EU in order to address influxes at an early stage and outside of the EU territory.

After the EU's "big bang" enlargement in 2004, the question of immigration – or, more accurately, the shape of a common immigration policy – was considered with renewed urgency. At issue, and of concern to "old" member states, was the movement of the EU's external border to new member states with "untested" border protection capabilities (Ruspini 2006). Internal security questions gained prominence, since the opening of internal borders was seen as a conduit for transnational crime, drug trafficking, and terrorism. This sense of urgency is clearly reflected in the Hague Programme (2004) and the Stockholm Programme's (2009) emphasis on managing borders.

A key part of the EU's border management approach is the creation of information databases and the use of new technologies (Dijstelbloem and Meijer 2011). For example, as a result of the creation of Schengen, police and judges can access information on individuals passing through the EU's external border. The Schengen Information System (SIS) compiles this data. An expanded system (SISII) will have the capability to store and exchange biometric data (Brady 2008: 24).[35] Information on visa applications within the Schengen area will be available via the Visa Information System (VIS). In operation since 2011, VIS contains photographs and fingerprint information from third-country nationals who apply for a Schengen visa (Ilies 2009). VIS complements EURODAC, a database designed to prevent abuse of member state asylum systems by communicating information regarding asylum applications in different member states.

More recently, a satellite-based European External Border Surveillance System (EUROSUR) was proposed to assist member states "in reaching full situational awareness" and to enhance "operational readiness" of law enforcement officials.[36]

Structures and venues

EU policies on immigration and asylum are proposed by the Commission (normally prepared therein by the former DG for Justice, Freedom and Security, now DG Home) and decided upon by the Council (acting in the form of the Justice and Home Affairs Council).

Frontex is one of the main structures through which EU member states cooperate on immigration issues (Leonard 2009, 2010). In May 2002, the Commission proposed an ambitious plan to create a "European Corps of Border Guards."[37] The plan, supported by EU member states bordering the Mediterranean, was roundly rejected. The Commission instead proposed a series of more modest steps toward "integrated border management" and waited for a more opportune time to create an agency (House of Lords 2008: 22). The European Council, meeting in Thessalonica in 2003, returned to the issue and invited the Commission to "explore the necessity of creating new institutional mechanisms, including the possible creation of a Community operational structure" for the management of external borders. In 2004, the Council agreed to its creation.[38] Frontex, seated in Warsaw, Poland, took up its operational responsibilities in 2005.

The regulation establishing Frontex sets out six tasks for the agency: (1) coordinate operational cooperation among member states in the field of management of external borders; (2) assist member states on training of national border guards, including the establishment of common training standards; (3) carry out risk analyses; (4) follow up on the development of research relevant for the control and surveillance of external borders; (5) assist member states in circumstances requiring increased technical and operational assistance at external borders; and (6) provide member states with the necessary support in organizing joint return operations.

While Frontex has received much attention, its actual remit is rather limited. Each EU member state remains responsible for managing its external borders, which means Frontex only can seek to further operational cooperation among member states. Since it has few "own resources," it relies on the resources (e.g. officers, ships, helicopters) of member states.

In 2007, EU interior ministers agreed to establish a rapidly deployable force of border guards to assist countries facing an immigration "emergency." RABIT teams, as these rapid border intervention teams are known, operate via Frontex and include a pool of over 450 national experts made available at short notice (within five working days) to any member state whose national borders are under "urgent and exceptional" strain from unexpected migration (Ilies 2009: 7). All member states must participate in RABIT, and once contributions are made available, each member state is obliged to mobilize those contributions when requested (House of Lords 2008: 40). Record high levels of immigration into the EU along the Greek–Turkish border in the second quarter of 2010 prompted the deployment of the first-ever RABIT team to help stem the flows and properly process asylum applications.

Managing epidemics

It was the Maastricht Treaty, signed in 1992, where member states first agreed to some cooperation in the area of public health. Article 219 confirmed that combating disease and enhancing public health should be an EU competence, although actions remained limited to research programs, education campaigns, and other "light" forms of coordination (Lezaun and Groenleer 2006).

In 1998, in the wake of the "mad cow" crisis, member states agreed to stronger cooperation in the form of a "communicable diseases network."[39] That network, which is still in operation today (operated by the European Centre for Disease Prevention and Control [ECDC]), is used for surveillance, early detection of diseases, and early communication of response measures.

Further crises, including the dioxin scandal (1999) and SARS (2002), highlighted European vulnerabilities to health threats (MacLehose, McKee, and Weinberg 2002). The emergence of H5N1 avian influenza in 2005 sparked a number of emergency meetings on the state of preparedness in Europe.[40] Ministers agreed that "EU member states need to coordinate efforts in the face of a risk of a human pandemic" and pledged to "ensure strong coordination and information sharing" to minimize uncertainties during a pandemic outbreak.[41]

Concerns surfaced regarding a lack of pandemic preparedness planning in the member states. A basic level of capacity to survey outbreaks, act on them, and adequately inform European partners was singled out as an especially important need. Better coordination among EU agencies such as the European Agency for the Evaluation of Medicinal Products (EMEA), the European Food Safety Agency (EFSA), and the ECDC was encouraged.

Much of the public discussion on the EU's response to H5N1 focused on preparatory action, especially the stockpiling of vaccines and antivirals. The Commission continuously pushed member states to abide by the WHO's recommendation of stockpiles for at least 25% of the population. Debate soon emerged as to the benefits of EU-level coordination and management of antiviral stocks, with Italy, Belgium, Lithuania, and Cyprus calling for centralized stockpiles and common stock management. Other countries, particularly France and the UK, balked at these proposals. Pharmaceutical companies based in Europe were particularly strident in their criticism of the EU states' lack of preparation and the long delays preceding their requests for vaccine manufacturing. Another issue causing controversy was related to the vaccination of birds. In February 2006, the Commission gave France and the Netherlands permission to vaccinate poultry stocks against H5N1, despite disagreement on the vaccine's effectiveness and concerns that vaccination would mask the outbreak of a disease.

Two additional initiatives deserve notice. First, the EU is authorized to place certain communicable diseases on a watch list for monitoring and tracking.[42] This list is updated when deemed necessary by the Commission, thus obligating member states to increase surveillance and reporting capacity regularly. The list is coordinated closely with the WHO. The addition of new diseases to the watch list has led to disease-specific networks (approximately 17) that further bind national health authorities together. Perhaps the most prominent of these networks is the European Influenza Surveillance Scheme that draws data from member state sources to create a weekly influenza report. Second, shared laboratory detection standards and common hospital treatment guidelines for communicable diseases have been created in an attempt to build capacities across the EU (Rhinard 2009).

In June 2009, the WHO declared H1N1 (also known as the "swine flu") a full pandemic, which reinforced the importance of these initiatives. The epidemic led to newfound attention to existing measures (many initiated after the outbreak of H5N1 "bird flu") along with a handful of new measures related to planning, situation monitoring and assessment, reducing the spread of the disease, providing adequate health care, and communicating with the public (Brattberg and Rhinard 2011).

Instruments and venues

Public health policy at the EU level is formulated and managed through the Commission's DG for Public Health and Consumer Protection (DG Health or known internally by its French acronym, "DG Sanco"). Implementation and day-to-day policy management of health threats is the task of the "health threats unit" founded in 2003 and located in Luxembourg. Here we find a Crisis Room in which health-related rapid alert systems are based (such as the Early Warning and Response System and the Rapid Alert System for Biological and Chemical Attacks and Threats, or "RAS-BICHAT").

In response to concerns that member states were not adequately prepared for communicable diseases, the ECDC was founded in 2004 in Stockholm, Sweden. The mission of the agency is to "identify, assess, and communicate current and emerging threats to human health from communicable diseases."[43] The ECDC is, in other words, intended to consolidate the surveillance and early warning responsibilities that have been delegated to the European level. The ECDC is also intended as a repository of scientific knowledge that can be deployed across Europe and as a source of "best practice" for countries that may not have sufficient capacity.

Managing critical infrastructures

Critical infrastructures reach across national borders.[44] A role for the EU in guarding against breakdowns and preparing for failures therefore would seem self-evident. Member states agree that the EU should have a role, but the precise nature of such a role remains unclear. Member states insist that critical infrastructure protection must remain a national competence and that European-level programs should show clear "added value" and remain voluntary. Very few initiatives, beyond a large number of meetings, conferences, and feasibility studies, have emerged on this issue since an initial spate of ideas and proposals.

Global attention to so-called critical infrastructures can be traced to the Y2K threat identified during the second half of the 1990s. Attention quickly dissipated after the feared paralysis of critical structures did not materialize in the first hours and days of the new millennium. After the events of 9/11, the protection of critical infrastructure again moved to the top of national security agendas (Auerswald *et al.* 2006).

In 2002 the EU first paid heed to the importance of critical infrastructure protection (CIP) as an explicit goal of policy coordination. In its program to "improve co-operation for preventing and limiting the consequences of CBRN terrorist threats," the Commission argued that the potential for catastrophic terrorist attacks that affect critical infrastructures was increasing in Europe.[45] The JHA Council encouraged the Commission to look further into ways to enhance critical infrastructure protection.

In June 2004, heads of state and government meeting in a European Council summit invited the Commission to prepare an overall strategy to protect the Union's critical infrastructures. The Commission responded in 2005 with a "Green Paper" laying out ideas for a "European Programme for Critical Infrastructure Protection" (EPCIP).[46]

The Green Paper suggested an ambitious goal for EPCIP: to ensure both adequate levels of protection for European critical infrastructures and rapid recovery arrangements. Noting that some degree of pan-European supervision might be necessary, the Green Paper routinely acknowledged the importance of subsidiarity. The EU role would concentrate only on aspects of CIP with cross-border effects. The Commission issued a package of initiatives, which comprised three elements:

- First, a proposal for a directive "on the identification and designation of European Critical Infrastructure and the assessment of the need to improve their protection." This proposal was seen as a modest first step on which an eventual program would be based.
- Second, a Communication on a "European Programme for Critical Infrastructure Protection" which presented the (few) agreed-upon findings from the consultation, including non-binding measures intended to encourage convergence in national protection measures, to ensure each member state designated a cross-sectoral

critical infrastructure protection official, and to set out the framework for an "EPCIP Action Plan" for future work.
- Third, a Communication outlining steps forward for critical infrastructure protection in two specific sectors (in which some degree of consensus has been reached): energy and transport. This Communication on "Protecting Europe's Critical Energy and Transport Infrastructure" represented the most progress made on critical infrastructure protection, arguably because it focused on a limited number of goals (Fritzon *et al.* 2007).

In April 2007, cyberattacks swamped websites of Estonian government agencies, communication services, media outlets, and major banks. The attacks forced Estonia to shut down all incoming foreign internet traffic. The scale of the attack, the organization of the attackers, and the threat it posed to Estonian national security took this incident beyond a minor hacking incident to a full-fledged internet sabotage with implications for regional security and stability. Cyber security subsequently assumed a prominent position in both the Stockholm Programme and the EU's Internal Security Strategy.

We can find additional and often more concrete initiatives aimed at protecting critical infrastructures by looking into specific policy areas (Fritzon *et al.* 2007). Within the food sector, for instance, European-level attention is being placed on protecting the food supply chain "from farm to fork" through standard setting, inspections and harmonization.[47] Legislative proposals have been formulated to maintain minimum stocks of crude oil and petroleum products, putting in place emergency procedures to be used in the event of a shortage.[48] Road, sea, and air transport policies in the EU have been supplemented with a variety of security measures intended to prevent breakdowns in critical transport networks. One focus is on tunnel security and transportation of dangerous goods, addressed by harmonizing minimum safety requirements and enhancing tunnel safety.[49]

The 2010 Icelandic ash crisis, which paralyzed Europe's air traffic, revealed the haphazard nature of European air traffic management.[50] Responsibility for opening and closing airspaces, planning routes, and grounding flights rested solely with individual European countries. European authorities lacked a central authority to provide critical scientific data and policy recommendations that could have clarified and expedited European response efforts.[51]

In the aftermath of the ash cloud crisis, the EU moved to improve air traffic coordination.[52] This initiative is a part of the Single European Sky initiative, an effort to reform the architecture of European air traffic control. To this end, the European Commission is working closely with the European Organisation for the Safety of Air Navigation (EUROCONTROL).[53]

Institutional structures

The EU's critical infrastructure protection policy is managed by officials from the Commission's DG for Home Affairs and decided on by the Council formation of Justice and Home Affairs ministers. One structure to emerge in this area is CIWIN, a critical infrastructure warning information network (proposed in the Hague Programme). CIWIN brings together member state CIP specialists to assist the Commission in drawing up programs to facilitate exchange of information on shared threats and vulnerabilities and appropriate countermeasures and strategies.

Within specific sectors, additional structures can be found. For the protection of critical information infrastructures, for instance, the EU's agency for network and information security (ENISA) offers a hub for coordination.[54] ENISA has a broad but fairly shallow mandate, charged with assisting member states, public organizations, and industries to prevent, manage, and solve problems of ICT security. ENISA, however, has no decision-making authority on policy questions. The Commission's *Internal Security Strategy in Action* announced the initiation of the EU Cybercrime Centre, partly as a response to the April 2007 cyber attacks against Estonia. The center became operational in 2013 and is placed within Europol.

The EU has launched a project focused on the resilience of trans-European energy networks. The TEN-E project offers structures in which member state authorities can gather to consider infrastructure supply and protection issues, with resources available for cross-border studies and more operational projects.[55] A similar project supports trans-European networks related to roads, rail, waterways, and air transport. In the area of aviation security, a European aviation safety agency (EASA) was established in 2002 with wide-ranging authority for regulatory and executive tasks (Pierre and Peters 2009). EASA provides technical expertise to the European Commission by assisting in the drafting of rules for aviation safety in various areas and providing technical input to the conclusion of the relevant international agreements.[56]

Managing financial crises

The global financial crisis brutally exposed the lack of transboundary crisis management capacities – sector specific and generic – at the European level. It underlines the crucial importance of developing transboundary crisis management capacities, as individual member states simply cannot deal with crises of this severity and scale. European leaders consistently failed to "get ahead" of the financial crisis, which, in turn, fed a legitimacy crisis for the euro and the very idea of European unity. The citizens of southern member states suffered from sky-high unemployment figures, receding government services, and related public order incidents.

The global financial crisis demonstrated, more specifically, two shortcomings of the EU's financial infrastructure (Quaglia 2013). First, it exposed the absence of financial supervision at the European level. Second, it showed that a lack of crisis management instruments translates into slow and incremental decision making. The EU has reacted with a spate of policy and institutional innovations (Antoniadis 2011). Whether these will be enough to save the euro and restore trust in the EU's economies remains to be seen.

Financial regulation and supervision

In a single market, where a financial institution authorized to provide financial services in one member state can provide the same services throughout the market, one would expect regulation and supervision to guarantee a "level playing field" for cross-border banking, insurance, and securities and investment funds. One would expect some sort of coordination of the more than 60 financial supervisors in European Union countries (Gualandri and Grasso 2006; House of Lords 2009b).

In 1985, the European Commission proposed a White Paper on the Internal Market. The "Lord Cockfield Paper" highlighted the need for a more efficient and integrated regulatory structure for European economies. Legally enshrined by the 1986 Single European Act, Lord

Cockfield's ideas relaunched the drive to complete the EU's internal market. The creation of a regulatory framework around a single market for financial services was part of that drive.

During the 1990s, a number of criticisms were lodged against EU financial legislation. Swiftly changing market conditions often made proposals out of date by the time they were approved. There were worries about the disconnect between EU-agreed laws and implementation effectiveness in national capitals. In response to these problems, heads of state and governments in July 2000 established a Commission of Wise Men on the Regulation of the European Securities Market under the chairmanship of Alexandre Lamfalussy (Gualandri and Grasso 2006). Its report recommended major reforms to speed up financial regulation, making it more flexible and connecting it more closely with national regulators. The Lamfalussy group's proposals were adopted by EU leaders in 2001.

The introduction of the euro provided a major impulse for new regulation aimed at establishing a single market in financial services. It was realized that use of the euro by financial institutions across borders would reveal a patchwork of national regulators and would challenge the abilities of national-based financial regulators and services.

To that end, the Commission identified five priority areas for completing the single market in financial services:[57] (1) creating a new legislative process for the supranational regulation of financial markets; (2) eliminating fragmentation in capital markets; (3) lowering any remaining obstacles to cross-border retail banking; (4) coordinating national supervisory authorities; and (5) facilitating cross-border financial transactions for consumers and banks.

These five areas, and the challenges they sought to overcome, provided the main components of the Commission's Financial Services Action Plan, a document that spawned numerous proposals and provided the backdrop for the EU's heightened role in financial regulation. It was complemented by a White Paper on Financial Services Policy, which outlined the Commission's financial services policy priorities through 2010.[58]

Although the EU's role in financial *regulation* grew during this period, its role in financial *supervision* did not.[59] Before the crisis, national supervision authorities cooperated in several specialized committees (the so-called Lamfalussy committees) and the Banking

Supervision Committee of the European System of Central Banks. Member states even agreed on several "high-level principles of cooperation" for crisis, which found their way into several Memoranda of Understanding (in 2003 and 2005). In the first phase of the crisis, the Economic and Finance Council (ECOFIN) agreed on additional principles which were incorporated in these Memoranda in 2008.

But national governments have been slow to give up substantial supervisory responsibilities to the European level, and there is only limited legal authority in the EU treaties on the subject of supervision. This disconnect became a source of intense debate during the financial crisis.

Managing the fallout of a global financial crisis

The global financial crisis followed the collapse of the US housing market (Financial Crisis Inquiry Commission 2011; Quaglia 2013). In the summer of 2007, the US subprime mortgage market imploded as homeowners struggled to make their payments when their interest rates rose and property prices fell. The problems in the housing market soon spread to the US banking sector and the international financial system at large (Hodson and Quaglia 2009).

Several European banks suffered large subprime-related losses. In August 2007, the German government rescued IKB Deutsche Industriebank. Some days later, BNP Paribas suspended three of its investment funds, claiming a complete lack of liquidity in certain market segments of the US securitization market. The European Central Bank (ECB), in an effort to ensure orderly conditions in the euro area markets, responded by offering €100 billion in short-term financing to banks. This would be the first in a series of emergency liquidity measures by the ECB and monetary authorities worldwide.

The collapse of the US investment giant Lehman Brothers on September 15, 2008, plunged financial markets into turmoil and led to worldwide destabilization. One EU member state after another was forced to rescue overextended banks.

The Commission argued that its role in the financial crisis was limited. It would apply EU law, notably on competition and state aid, to maintain a level playing field. In addition, it would work "behind the scenes, as a 'bridge', facilitating agreement and ensuring coherence between national action and European action; between member states within the euro zone and outside it; and between all the institutions of the EU and the member states." The Commission emphasized that only member states have the legal authority and the capital to commit taxpayer money for "rescue plans" and to intervene directly in individual financial institutions.[60]

The European Central Bank played an increasingly crucial role in this crisis. The ECB, which defines monetary policy in the Eurozone (for more on the ECB, see below), cut the base rate to historic lows. It allowed banks to use a wider range of assets as collateral when borrowing from the Eurosystem and provided banks with unlimited liquidity at very low interest rates. It engaged in quantitative easing, by purchasing covered bonds and debt securities, backed by mortgages or public-sector loans, in order to ease credit conditions in the euro area (Hodson and Quaglia 2009).

In December 2009, following worse-than-expected reports on the Greek economy, the world's top credit ratings agencies downgraded Greece's sovereign debt. Greek officials announced sweeping spending cuts to bring the public deficit, which had ballooned to 12.7% of GDP, under control. But those announcements did little to soothe the nerves of either financial markets, which continued to sell off Greek bonds, or those of the public, which responded in February 2010 with mass protests in the streets of Athens.

On March 25, 2010, guided in the negotiations by Herman van Rompuy, the then-newly appointed President of the European Council, Eurozone leaders agreed to offer the Greek government a three-year financial aid program amounting to €110 billion. This marked the beginning of a series of support actions and rescue packages for Greece.[61]

On May 9, 2010, the EU member states agreed to create the European Financial Stability Facility (EFSF), aimed at ensuring financial stability in Europe through the provision of financial assistance to Eurozone countries in need of economic assistance. The EFSF was backed by guarantee commitments from the Eurozone member states for a total of €780 billion and has a lending capacity of €440 billion.[62]

The financial crisis in Europe took yet another turn when Ireland ran into financial trouble. In November 2010, the EU agreed on an €85 billion rescue deal for Ireland. Three initial steps were taken by governments to expand the EU's role in battling future crises. First, a stricter Stability and Growth Pact (which sets rules over sovereign debt in the Eurozone) was proposed, one which contained more sanctions and binding power over governments. Second, a new economic surveillance mechanism was introduced. Third, a European Stability Mechanism (ESM) was adopted to replace the temporary European Financial Stability Facility (EFSF) in mid-2013.

In April 2011, Portugal confirmed it would need a bailout (to the tune of €80 billion). In the summer of 2011, Italy and Spain began to tread water. During the EU Summit on October 26–27, 2011, EU leaders agreed to increase the EFSF from €500 billion to €1 trillion. In June 2012, the EU made available €100 billion for Spain to assist its banks. At the time of writing, it was not clear whether Italy and Spain would need additional assistance (and whether the EU would be able to provide it).

While the EU limped from one crisis to another, it did implement wide-ranging reform of the EU framework for financial regulation and supervision. The EU has moved to regulate Credit Rating Agencies, Alternative Investment Funds, and Over the Counter Derivatives. It has adopted policies that many considered unacceptable a few months before. And, as we will see below, it has created new institutions. In these reform efforts, improvisation reigned supreme. As the *Financial Times* commentator Martin Wolf put it, "the Eurozone's aeroplane is being redesigned while crashing" (Wolf 2012).

Several obstacles have stood in the way of a strong EU role in financial crisis management. One is that decision processes for ensuring financial stability and crisis management remain at the national level (Nieto and Schinasi 2008). Many of the Commission's proposals for greater coordination have met with reluctance or outright hostility (Evans-Pritchard 2009).

An additional factor is that the most important European players find it hard to agree on the type of capacities needed to contain financial crises. The key players in this discussion, German Chancellor Angela Merkel and French President Nicolas Sarkozy (later François Hollande), did not share the same view on an adequate response. The German government has consistently called for a stricter fiscal rulebook for Eurozone members (see also Boin, 't Hart, and van Esch 2012). The French government, on the other hand, has called for the formation of an "economic government" for the EU to promote stronger coordination of economic policy and to shrink the governance gap in Europe between monetary policy (handled by the ECB) and economic policies (handled mainly by member states).

During the crisis, the German government gradually shifted to stronger economic policy coordination at the EU level, more resources for the fiscal rescue mechanisms, and allowing the ECB to buy sovereign bonds. By early 2011, the German and French governments, after lengthy negotiations, reached an unexpected agreement by which they would support each other's proposals for more economic integration (in the area of pensions, wages, and labor rules) and stricter limits on government debt. Whether other EU member governments (and other Eurozone members, more specifically) will agree to such groundbreaking steps was an open question at the time of writing.

Institutional structures

The key actors involved in financial regulation and supervision in the EU include the European Commission (more specifically, DG Internal Market), the Council of the European Union (more specifically, the ECOFIN), and the European Central Bank.

The ECB became one of the most important crisis actors during the crisis. Headquartered in Frankfurt, the ECB serves as the central bank of the Eurozone. Like any central bank, its primary objective is to maintain price stability. Its key tasks are to define and implement the Eurozone's monetary policy, to conduct foreign exchange operations, to manage the foreign reserves of the European System of Central Banks, to promote a sound financial market infrastructure in Europe, and to authorize the issuance of euro bank notes (Leino 2000). The ECB's role as a lender of last resort is highly constrained, however (Schinasi and Teixeira 2006). It cannot recapitalize banks or apply massive economic stimulus measures.

Four additional institutions emerged in 2011. The European systemic risk board, chaired by the ECB president, will have powers to issue warnings and recommendations when it sees threats to economies or financial systems. This will help the ECB's capacity of surveillance.[63] In addition, the EU created the European Banking Authority (EBA), the European Insurance and Occupational Pensions Authority, and the European Securities and Markets Authority.[64]

Analysis: a fragmented field inching toward institutionalization?

This chapter has documented a wide variety of organizational and policy tools that can be – and periodically are – used to facilitate a joint response to transboundary crises. The list of tools presented here is certainly not exhaustive: in virtually every Commission DG, Council working group, or EU agency one can detect formal and informal capacities to bring member states together and make urgent yet deliberate decisions on crisis-related topics.

The field mapped in this chapter ranks relatively low on the institutionalization scale set out in Chapter 1 (certainly when compared to the fields described in Chapters 2 and 3). The capacities are scattered across the EU's civil service and there is no observable logic tying them together (cf. Kaunert, Leonard, and Pawlak 2012). There is no centralized department for transboundary crisis management; it is a field without a name (in organizational diagrams and in textbooks, this field does not feature). It is not even clear who in the EU is aware of all these available capacities.[65]

With the analytical lens of transboundary crisis management, however, there is much to see. What is striking about this emerging field, with all its different pockets, is the variety of institutional processes that contribute to its fragmentation. Some policies and organizational units have become fairly well established and appear to be ascending the institutionalization ladder (think of Argus, Frontex, Europol, and the ECB). Other initiatives, such as the ISS, are young but promising. Yet others are slow to take off and struggle to develop a recognizable identity (CCA and COSI). The different levels of institutionalization in this field explain why the field as a whole is not highly institutionalized.[66]

Why does the field of transboundary crisis management not rank much higher in terms of institutionalization? Not only is there a significant amount of instruments, structures, and venues, but the very idea of transboundary crisis management fits the core principles of subsidiarity and "adding value" (the idea that the EU should only initiate activities that member states cannot or will not do alone). Moreover, the EU has faced several transboundary crises that demonstrated why a joint response is both necessary and helpful. So why has this field not developed like those described in Chapters 2 and 3 – policy fields that arguably had less "institutionalization potential"? To explore this puzzle of under-institutionalization, we return to our institutionalization model as a heuristic device.

Ambiguous ambitions

Policy ambitions in this field are often high but tend to be rather ambiguous on closer inspection. There is, of course, ample reason to formulate sky-high ambitions for the EU's transboundary crisis management capacities. The continuing drive to create an open market across the European continent reliably produces opportunities for small problems to propel themselves across borders and snowball into a new, hard-to-manage type of adversity. Whether we speak of the Iceland ash crisis or the mysterious EHEC outbreak, the financial crisis or the unfolding consequences of climate change, it is widely realized that Europe faces the prospect of transboundary crises. It is, therefore, understandable and commendable to hear Cecilia Malmström, Commissioner for Home Affairs, argue that the EU should "better protect the life and safety of EU citizens against all threats and hazards."[67]

It is hard to find a clarion call for the creation of transboundary crisis management capacity at the EU level, however. Lofty rhetoric about a "secure Europe" typically suggests an important role for the EU. But references to the actual construction or enhancement of transboundary crisis management capacity tend to be vague, underdeveloped, or unfeasible. There is not even a definition of what would constitute a crisis for the EU.

One reason for this understated response to a recognized problem are differences in opinion among national governments over what the EU's role in transboundary crisis management should be. This is not just a problem of conflicting national interests (although that is often claimed as the main problem; see Monar 2010). It also reflects a lack of understanding about the essential characteristics of transboundary crises and the required response capacities at the European level. How do we prevent a smallpox outbreak or a cyberattack? What response capacities are needed when energy infrastructures shatter? In the absence of such knowledge, political leaders (especially national ones) are prone to cast the EU's role in ambiguous and modest terms: the EU should do more, but member states should retain a final say. This explains why the EU's formal competences rarely correspond with the kinds of activities required for managing transboundary crises.

Another reason is the linkage between security and crisis: the few concerted efforts to create transboundary crisis management capacities are mostly wrapped up in wider initiatives related to internal security. By formulating crisis management initiatives in terms of security, these initiatives fall within the domain of interior and justice ministers – not considered one of the EU's most reform-minded group of policy actors. It is hard for any ambitious initiative to survive contact with this arena (Brady 2010). Moreover, since the Lisbon Treaty has brought the Parliament and European Court of Justice into the highly sensitive areas of criminal justice and police cooperation, an atmosphere of reluctance and skepticism has descended on the internal security policy arena (Bossong and Rhinard 2013).

The absence of a clearly recognized "problem owner" does not help. The organization of the Commission into policy-specific directorates has long been the source of policy fragmentation and coordination problems (Christiansen 2006). For many years, this meant that the Commission's various crisis management initiatives, in the areas of transport or environment, for example, took place with little mutual awareness and even less overarching guidance. While DG Home has taken the lead in attempting to formulate a more comprehensive view on transboundary crisis management, it holds little or no authority over some of the relevant tools and policies. DG Home has declared its interest to "decide how we can implement the Solidarity Clause in a much better way," for instance, but the Civil Protection Mechanism (a tool that is pertinent to the Solidarity Clause) falls within the remit of DG ECHO.[68]

Under these conditions, it is no surprise that institutionalization in this area has been slow and uneven. It takes time, experimentation, and adaptation to translate ambitious and ambiguous policy aims into relevant, effective, and legitimate practices. Once initiatives get off the ground, they can grow by demonstrating their added value.

The question is whether this policy arena will have enough time to allow such institutionalization processes to occur. Crises and disasters are not likely to abate nor are they likely to become more "localized" and less transboundary. When they occur, they trigger repeated interventions by political leadership, which will result in additional and ever more ambitious policy aims. That will put the burden on the EU, which is then left to meet these ambitions. Institutionalization, in other words, will demand adaptive capacity on the part of EU institutions.

The capacity to adapt

We have seen how national leaders, when they meet in Brussels, are keen to profess the importance of transboundary crisis management, but member states remain wary of giving up too much of their own policymaking autonomy in exchange for supranational coordination. The contrast between enthusiastic political declarations and hesitant implementation has resulted in patchy capacities at the EU level. At the same time, and somewhat curiously, the EU policy process inexorably muddles on and continues to produce new transboundary crisis management capacities.

Adaptation here makes up for a lack of member state drive. The EU manages to translate highly ambiguous and seemingly impossible ambitions into practices that prove useful. The adaptive capacity we found in the previous chapters can be found in this field as well.

We can observe a pattern here. Most new policy initiatives on transboundary crisis management are reactions to systemic "shocks" (cf. Sabatier and Jenkins-Smith 1993; Argomaniz 2009). When such a crisis confronts member states with their shortcomings, the EU is quick to suggest a role for itself. In response to mad cow disease, for instance, member states agreed to new health security measures and the Commission built institutions for regulating animal and human health questions. More recently, terror attacks have had a galvanizing effect on collective action in crisis management (Argomaniz 2009). These events suggested a Europe-wide response, generating a host of new measures regarding internal security such as biosafety preparedness, customs controls, joint policing, and intelligence sharing.

Something then has to happen "on the ground" if these policy initiatives are to produce results. Perhaps the most typical example of how this tends to happen in this policy area is found in the institutional history of Europol. Initially distrusted widely and starting its life with clipped wings, Europol managed to grow step by step. Rob Wainwright, director of Europol, explains how this process works:

> From the start, we have had to fight for our natural space in the hearts and minds of the police. We have to do that through hard labor, demonstrating

> to investigators in our 12,000 to 13,000 cases the unique value that we can add . . . One of the cornerstones of our ambition is to become what I would describe as the information hub for police work in the EU. (House of Lords, 2011b: 48).

The financial crisis tested the EU's capacity to adapt "in real time." While the EU has been consistently criticized for not "getting ahead" of this crisis, it also demonstrated a remarkable potential to revise policies and initiate new tools "on the fly" – ushering in reform that was considered unthinkable only months earlier.

This is, of course, not true for each initiative launched in this policy area. One of the most promising tools – the Crisis Coordination Arrangements – was widely lauded but has been used only sparingly. Other organizations such as Eurojust and ENISA struggle to gain widespread recognition. The EU Counterterrorism Coordinator appears to be in an "existential crisis" (since that position's mandate is narrow and not particularly clear). The COSI was off to a slow start and has shown few signs of establishing itself.

Surprising legitimacy: a potential source of forward momentum

Even though the formulation of ambitions and the coherence of EU action in this area reveal a considerable amount of uncertainty and fragmentation, we see pockets of growing legitimacy for the EU's transboundary crisis management capacities. Europe's citizens overwhelmingly agree that the EU should work to make Europe safer. Even when the issue becomes more concrete and, thus, potentially contentious, such as in the field of criminal justice and border management, we see that initiatives hold up quite well in the court of public opinion. Opinion polls confirm that the public sees cooperation on terrorism, health threats, cybercrime, and border security as a key role for the EU.[69] Organizations such as Europol and Frontex continue to gradually expand their remit and activities without provoking public debate.

The European Parliament, while occasionally skeptical of security-related initiatives, generally approves of the EU's broader role of acting to safeguard the well-being of European citizens. This is apparent in political declarations, in which the EP claims credit for assisting in such efforts as well in evaluations of the EU's role in handling crises.[70]

Initiatives to enhance transboundary crisis management capacities rarely evoke the ire of member states. Ministers often agree to upgrade capacities after crises. Indeed, rarely does a meeting of government officials (in the EU's Council of Ministers, for instance) pass without reference to the need for stronger cooperation to overcome the limitations of national action. One (perhaps surprising) example is the UK. A House of Lords inquiry (2011a: 11) on the Internal Security Strategy concludes that "national security of a State is the responsibility of that State but it cannot be dealt with by that State acting alone."

All this does not mean that member states are ardent cheerleaders for an expanding EU role. Most governments remain hesitant at best about using some of the EU's capacities. To some extent this is a matter of national pride, in that some countries do not want to appear "weak" and in need of assistance. This may also be a matter of sovereignty, because some countries may not want to share what they feel to be sensitive information. While this hesitation is understandable and unlikely to dissipate any time soon, it explains why we cannot expect rapid or deep institutionalization of transboundary crisis management capacities at the European level – even if it would be a rational and desirable course of action in the light of future crises.

Conclusion: the incremental emergence of transboundary crisis management capacity

One might argue that the EU was born out of the biggest transboundary crisis ever witnessed on the European continent. On the ruins of war-ravaged Europe, six nations came together to create a unique form of cooperation to make sure a similar disaster would never occur again. The founders of the European Community built strong economic ties to help member states rebuild and prosper, while preventing future manifestations of the distinctly transboundary threat of war.

In its 50-plus years, the EU has not seen any wars within its boundaries. But it has confronted several transboundary threats, which have posed complex challenges to member states and the EU. The Chernobyl disaster (1986), the BSE crisis (1996), the terror attacks on Madrid (2004) and London (2005), the cyber attacks in Estonia (2007), the H5N1 pandemic (2005), the Iceland ash crisis (2010), and the unfolding financial crisis posed common threats, requiring a shared response.

In response, the EU has developed transboundary crisis management capacities: instruments, policies, and venues that can be used in response to transboundary threats and crises. But these capacities have not, or hardly, been institutionalized. This chapter suggests that the transboundary capacities we found in the various policy sectors are in a nascent stage of institutionalization.

The emergence of new transboundary threats and crises will set the stage for further institutionalization, along the lines suggested in our first chapter. This can happen quite quickly, as demonstrated in the financial realm. It can move slowly, as seems to be happening in the area of critical infrastructure protection. As it stands, the EU is still far removed from acting in an autonomous fashion in response to transboundary crises. Member states maintain full authority and choose which tasks they want to delegate to the EU. Many member states are also slow to implement commitments made in EU venues (Crosbie 2007).

What is most needed, perhaps, is the development of integrated (e.g. non-sector specific) transboundary crisis management capacities such as CCA. This chapter demonstrates that such capacities are still in an early phase of institutionalization. They remain unused and, thus, untested, which explains the lack of institutionalization (in terms of our theoretical framework). But some EU capacities continue to grow, as evidenced by the proposed consolidation of intelligence venues into a Situation Awareness Center or the regular meetings of crisis-related administrators in the Commission. This could well prove to be a crucial step on the institutionalization ladder.

Over the years, the EU has gotten much more powerful in this area of transboundary threats. European leaders remain committed to the ideal of an integrated Europe and continue to endow the Union with more powers. While there is an urgent need for these growing capacities, it has become less and less clear what the EU is actually going to do with these new powers (Brady 2010). We will offer some ideas of our own in the next chapter.

Notes

1 One could argue that the oil crisis of the early 1970s was the first real transboundary crisis that the then small group of member states faced

together. It gave rise to regular Council meetings on the issue. Regular Council meetings have since become institutionalized.

2 In the early 1980s, the IAEA had guidelines drawn up for "mutual emergency assistance arrangements in connection with a nuclear accident or radiological emergency" (Medvedev 1992).

3 While the European Commission has been operating a rapid alert system for food since 1979, the RASFF was created in 2002 by the EU Food Legislation (regulation 178/2002, Chapter IV). In essence, RASFF was created to allow authorities in Europe to exchange information with each other in order to improve a rapid and coordinated member state response to health threats caused by food or feed. According to RASFF's annual report, in 2010 the number of notifications amounted to over 8000, representing an 8% increase from the previous year. In 2011, RASFF played an important function during the EU's response to the E. coli crisis. According to EU Health Commissioner Androulla Vassiliou, RASFF is "one of the great success stories of the EU's integrated approach to food safety, using to maximum effect the power of communication and collaboration."

4 We note that some authors now refer to the EU's policy domain of Justice and Home Affairs as the EU's "Area of Freedom, Security and Justice." The latter refers to a legal concept introduced in the Amsterdam Treaty (1997) and reflects a policymaking objective (for a discussion, see Walker 2004). We stick with the term Justice and Home Affairs (JHA) for ease of use.

5 The official text giving rise to Argus: "It is appropriate for the Commission to establish a general rapid alert system called Argus, in order to enhance its capacity to react quickly, efficiently, and in a coordinated manner, in its domain of competences, to crises of a multisectoral nature covering several policy areas and that require action at the Community level, *whatever their cause*." "Commission Decision of December 23, 2005 amending its internal Rules of Procedure." 2006/25/EC, Euratom. L 19/20. Brussels: January 24, 2006. (Italics added.)

6 "The Hague Programme: strengthening freedom, security and justice in the European Union." 16054/04. Brussels: December 13, 2004, p. 24.

7 The typical examples used to justify this new arrangement were the possibility of needing to shut down airports, respond to a string of terror attacks, or manage the knock-on effects of a pandemic (Gustenau 2006).

8 Council of the European Union (2007). "Report and revised Manual on EU emergency and crisis coordination – Endorsement." 10011/1/07. Brussels: June 20, 2007. See also Larsson (2009).

9 A 2012 review proposed the development of capacity for "Integrated Situational Awareness and Analysis" as a support structure related to the CCA.

10 "Communication from the Commission to the European Parliament, the Council, the European Economic and Social Committee and the Committee of the Regions – Delivering an area of freedom, security and justice for Europe's citizens: Action Plan Implementing the Stockholm Programme." COM (2010) 171 final. Brussels: April 20, 2010.

11 Many of the threats mentioned in the ISS overlap with those outlined in the ESS.

12 Draft Internal Security Strategy for the European Union: "Towards a European Security Model", 7120/10. Brussels: March 8, 2010.

13 European Commission (2010). "Communication on the EU Internal Security Strategy in Action: Five Steps Towards a More Secure Europe." COM (2010) 673 final. Brussels: November 22, 2010.

14 DG Justice, Freedom, and Security was split in 2010, following a Commission reorganization, into two separate directorates-general: DG Home, responsible for internal police, immigration, and security questions, and DG Justice and Fundamental Rights, responsible for judicial cooperation.

15 Proposal for establishing the instrument for financial support for police cooperation, preventing and combating crime, and crisis management. COM (2011) 753 final. Brussels: November 15, 2011.

16 In addition, there is funding for various EU agencies such as Europol, Frontex, the European police College (CEPOL) and the European Monitoring Centre for Drugs and Drug Addiction (EMCDDA).

17 See Groenleer (2013) for a discussion of the role that EU agencies can and do play in the management of crises.

18 TREVI stands for Terrorism, Radicalisation, Extremism, and International Violence.

19 The EU defines a terrorist group as a "structured organization . . . of more than two persons, acting in concert to commit terrorist offences." Terrorism is interpreted as "an intentional act, which, given its nature or its context, may seriously damage a country or an international organisation, as defined as an offence under national law, where committed with the aim of: (i) seriously intimidating a population; (ii) unduly compelling a government or an international organisation to perform or abstain from performing any act; or (iii) seriously destabilising or destroying the fundamental political, constitutional, economic or social structures of a country or an international organisation." "Council Common Position of 27 December 2001 on the application of specific measures to combat

terrorism." 2001/931/CFSP (OJ L 344)93), December 28, 2001. See Art. 1(3).

20 As is often the case when the EU makes on-the-fly policy announcements, the instruments put in place were drawn from existing ones across EU policy sectors, including intelligence collection, critical infrastructure protection, regulations to prohibit the accumulation of material used in explosives, and EU–UN dialogue processes for terrorism issues.

21 "José Manuel Barroso, President of the European Commission: Fighting terrorism together in the EU, International Summit on Democracy, Terrorism, and Security, Madrid, 10 March 2005." Speech, March 11, 2005.

22 European Council (2004). "Declaration on combating terrorism." Brussels: March 25, 2004.

23 Council of the European Union (2004). "EU Plan of Action on Combating Terrorism." 10586/04. Brussels: June 15, 2004.

24 Council of the European Union (2002). "Council Framework Decision of 13 June 2002 on the European Arrest Warrant and the Surrender Procedures Between Member States." Pub. L. No. L 190/1 (2002). Brussels: July 18, 2002.

25 The Hague Programme established the "principle of availability," which prescribed that law enforcement officials in any state can request, and should receive, information necessary for crime fighting if that information is available (Dittrich 2005: 17).

26 Being on the list meant one could be subjected to freezing of funds and economic resources. European Union, "Countering terrorism Council updates list of terrorist organisations." European Council (2004). "Countering terrorism – Council updates list of terrorist organisations. PRES/04/92, 7517/04 (Presse 92). Brussels: April 5, 2004.

27 Council of the European Union (2005). Press Release 14390/05 (Presse 296). Brussels: December 1–2, 2005. For an overview of EU anti-terror measures, see Cottey (2007: 178–9).

28 The SitCen is part of the EEAS (see Chapter 3). Since it was moved to the EEAS in 2010, it has been reorganized. At the time of writing, it was not clear how the SitCen (or its constituent parts) would be incorporated into the emerging EEAS structure.

29 This policy area deals with the more problematic aspects of immigration. It can be difficult to distinguish between illegal immigrants and asylum seekers, since many use the same routes and strategies for getting into the EU and make similar claims to legitimize their movement. In the light of this book's topic, and considering limitations of time and space, our focus here will be on policies and initiatives taken in the EU to prevent and manage mass influxes of migrants. As an aside, we note that the

EU still lacks a genuine "European Asylum Policy" and makes progress in only a piecemeal fashion due to member state reluctance to share sovereignty on this core responsibility of national governments (Collett 2010).

30 The Lisbon Treaty has changed the decision-making procedure for all immigration and asylum issues, from varying degrees of unanimity to complete qualified majority for all policies.

31 European Council (1999). Presidency Conclusions, Item 11, Tampere European Council October 1999.

32 European Council. "Plan for the management of the external borders of the Member States of the European Union." 10019/02. Brussels: June 14, 2002.

33 For the most part, these standards are now consolidated in the "Schengen Borders Code," which sets rules for assessing individuals at borders and for managing border traffic. "[E]stablishing a Community Code on the rules governing the movement of persons across borders" (Schengen Borders Code). Council Regulation 562/2006). Brussels: April 13, 2006.

34 "Communication on the Managed Entry in the EU of Persons in Need of International Protection." COM (2004) 410 final.

35 SISII, following a number of technical problems delaying its implementation, is currently planned to go online in the first quarter of 2013. "Regulation (EU) No 541/2010 of 3 June 2010 amending Regulation (EC) No 1104/2008 on migration from the Schengen Information System (SIS 1+) to the second generation Schengen Information System (SIS II)." Brussels: June 22, 2010.

36 EUROSUR entered a pilot phase in 2011 and should be fully operational by 2013.

37 "Communication from the Commission to Council and the European Parliament: Toward Integrated Management of the External Borders of the Member States of the European Union." COM (2002) 233. Brussels: May 7, 2002.

38 Relevant documents are: Council of the European Union (2003). "Proposal for a Council Regulation establishing a European Agency for the Management of Operational Co-operation at the External Borders." COM (2003) 0687 final – CNS 2003/0273 Brussels: November 20, 2003; Council of the European Union (2004). "Council Regulation (EC) No. 2007/2004 of October 26, 2004 establishing a European Agency for the Management of Operational Cooperation at the External Borders of the Member States of the European Union."

39 "Decision No 2119/98/EC of the European Parliament and of the Council of 24 September 1998 setting up a network for the epidemiological surveillance and control of communicable diseases in the Community."

40 The H5N1 virus was first detected in Hong Kong in 1997, affecting both wild birds and domestic poultry. During the summer of 2005, the capacity of H5N1 to transmit from animals to humans became apparent (Coulombier and Ekdahl 2005).

41 Informal Meeting of EU Health Ministers: "Pandemic Flu Preparedness." Press Release, Council of Ministers. Brussels: October 20–21, 2005.

42 European Commission (2007). "Decision amending Decision No 2119/98/EC of the European Parliament and of the Council and Decision 2000/96/EC as regards communicable diseases listed in those decisions." 2007/875/ EC. Brussels: December 18, 2007.

43 "Regulation (EC) No 851/2004 of the European Parliament and of the Council Establishing a European Centre for Disease Prevention and Control." L 142/1, Brussels: April 30, 2004.

44 Critical infrastructures are defined as "those physical and information technology facilities, networks, services, and assets which if disrupted or destroyed, would have a serious impact on the health, safety, security, or economic well-being of citizens or the effective functioning of governments in the member states." See also "Communication from the Commission to the Council and the European Parliament – Critical Infrastructure Protection in the fight against terrorism." COM (2004) 0702 final. Brussels: October 20, 2004.

45 Council of the European Union (2002). "Inventory of the EU instruments relevant for the Council and Commission Programme of 20 December 2002 to improve cooperation in the European Union for preventing and limiting the consequences of chemical, biological, radiological, or nuclear terrorist threats." 15873/02. Brussels: December 20, 2002.

46 Commission of the European Communities. (2005). "Green Paper on a European Programme for Critical Infrastructure Protection." COM (2005) 576 final. Brussels: November 17, 2005.

47 European Commission (2008). *Food: From Farm to Fork Statistics 2008*. Eurostat Pocketbooks.

48 Council of the European Union. (2009). "Imposing an obligation on Member States to maintain minimum stocks of crude oil and/or petroleum products." No 2009/119/EC. *Official Journal of the European Union* L 265/9. Brussels: 9 October 2009.

49 Directive 2004/54/EC of the European Parliament and of the Council of April 29, 2004, on minimum safety requirements for tunnels in the Trans-European Road Network.

50 The eruption of a volcano near the Eyjafjallajökull glacier on Iceland on April 14, 2010, created a huge ash plume, which caused the closure of airspace over much of European airspace. Some 10 million passengers were affected. Experts estimate the airline industry's costs to be €1.3

billion. The total losses due to the ash cloud were estimated by one study to be close to €5 billion. See Oxford Economics (2010) and Alemanno (2011).

51 European countries relied heavily on information based on a computer simulation from the Volcanic Ash Advisory Centre, located in the UK.

52 The EU Commission has proposed establishing a system of Functional Airspace Blocks to be implemented by December 2012. These blocks are intended to contribute to defragmenting the European airspace and to allow for economies of scale by allowing more direct flight routes.

53 EuroControl is a European organization with the goal of creating a single, unified European air traffic management network and is actively involved in Single Sky initiatives.

54 ENISA was created in March 2004 as part of the implementation of the EU's policies on information communication technology. "Regulation (EC) No 460/2004 of the European Parliament and of the Council of 10 March 2004 establishing the European Network and Information Security Agency (Text with EEA relevance)." Brussels: March 13, 2004.

55 "Proposal for a Regulation of the European Parliament and of the Council on guidelines for trans-European energy infrastructure and repealing Decision No 1364/2006/EC." Pub. L. No. COM (2011) 658. Brussels: October 19, 2011.

56 "Regulation (EC) No 1592/2002 of the European Parliament and of the Council of 15 July 2002 on common rules in the field of civil aviation and establishing a European Aviation Safety Agency." Brussels: September 7, 2002.

57 European Commission (1999). "Financial Services: Implementing the Framework for financial markets: Action Plan." COM (1999) 232, Brussels: May 11, 1999.

58 European Commission (2005). "White Paper: Financial Services Policy 2005–2010." COM (2005) 629 final. Brussels: December 1, 2005.

59 The supervisory aspect of financial markets includes ensuring that "banks or financial institutions subject to regulation follow the rules correctly and uniformly, that they adequately manage their risks, and that they adhere to certain minimum standards" (House of Lords 2009b).

60 European Commission (2008). "FAQs on Europe's response to the Financial Crisis." MEMO/08/618. Brussels: October 14, 2008.

61 Eurozone leaders agreed on a second package on July 21, 2011. On February 21, 2012, Eurozone leaders agreed on a third package, totaling an additional €100 billion in loans to Greece (Antoniadis 2011).

62 http://www.efsf.europa.eu/about/index.htm (accessed: January 15, 2012).

63 Ralph Atkins (2011) EU authority to head off economic crises. *Financial Times*, January 4.

64 The head of the EBA, Andrea Enria, warned that he plans to use the "true power" of a single set of rules to impose more uniform oversight on banks. In emergencies the board can overrule national regulators (*Financial Times*, February 15, 2011). In July 2011, the EBA released "stress tests" of 91 banks in 23 countries. These stress tests were more credible than the 2010 tests administered by the Committee of European Banking Supervisors (the EBA's predecessor). These earlier stress tests were discredited within months: by the end of the year Irish banks, which had passed the test, had to be bailed out.

65 In 2005, around the time we started the research project on which this book is based, a common response from interviewees questioned about transboundary crises was "we don't do that kind of thing." In 2009, the Secretariat-General of the Commission made a first attempt to create an overview of available capacities. The result, titled "Inventory of crisis management capacities in the European Commission and Community Agencies," was published as an internal Commission document and succeeded in raising awareness within the Commission of the myriad of capacities across its services. It also allowed us the opportunity to verify our data against that collected officially. In January 2011, a "corrigendum" was produced.

66 See Boin (2001) on the relation between field-level institutionalization and the institutionalization of its parts.

67 Speech at the Conference on research development and innovation for a more secure Europe. Oostende, November 23, 2010. This ambition is, of course, overstated. The ISS aims to define "a few limited but common threats."

68 Commissioner Malmström in her testimony to the House of Lords (2011b: 3).

69 See Eurobarometer (2006). *European Commission Presents Results of Eurobarometer on Avian Flu*. Eurobarometer (2011) *Special Eurobarometer 371: Internal Security*.

70 European Parliament (2012). "Strengthen Internal Security while Protecting Fundamental Rights, say MEPs." Press Release, Plenary Session, May 22, 2012. Reference No. 20120522IPR45545. See also European Parliament (2011). "Report on the EU Counter-Terrorism Policy: Main achievements and future challenges." Committee on Civil Liberties, Justice, and Home Affairs, July 20, 2011. Reference No. A7–0286/2011.

5 *Managing future crises*

Challenges and prospects for the European Union

The watchwords ought to be: maximizing synergies and avoiding "hard" or artificial splits between how we handle EU internal and external crises.

(Baroness Ashton 2010)[1]

The lesson of the past is that this is an area in which breakthroughs are often less than they seem, and in which the slow evolution of policy is more important than treaty provisions, declarations or individual agency.

(Smith 2003: 563)

From "mapping" to assessing EU crisis management capacities

This book documents the emergence of the European Union as a multifaceted crisis manager. It describes the many crisis management capacities now available at the supranational level in Europe, including unique forms of cooperation, surprisingly effective decision procedures, organizations with special competences, and even a degree of operational capabilities.

It is true that these capacities are scattered across the EU's institutional landscape. We had to search for some, as they were not designed with crisis management in mind. Some capacities are brand new; others are tried and tested. It is also true that some capacities have worked better than others.

Taken together, we can say that the EU has developed a wide range of tools that enable the Union to play a role in a joint response to a variety of crises. The combined resources of 27 member states and the distinct competence the EU has developed over the years have made the EU a unique actor in the international crisis arena.

The European Union applies its newfound capacities actively and visibly in a wide range of crises and disasters. The EU assists in the aftermath of disasters across the world, it directs civil–military missions to hot spots in the EU's neighborhood and beyond, and it coordinates member state actions in response to so-called transboundary threats. Few would have predicted this development just 15 years ago.[2]

This development of crisis management capacities has not come a moment too soon. More and more crises are arriving at the Union's door demanding attention: the continuing threat of pandemic outbreaks, the collapse of the financial system, infrastructural vulnerabilities, shifts in the global power constellation, new forms of terrorism, cyber warfare, and climate change loom on the horizon. These threats demand a robust, effective, and transparent system of crisis management.

The EU does not have such a system quite yet. It often appears "behind the curve" when unforeseen crises emerge. The financial crisis, the volcanic ash crisis, the Haiti earthquake disaster, and the recent upheaval in the Middle East surprised and divided the Union in a rather familiar way. While much has been accomplished in recent years and more capacity development is in the offing, the question is whether it will be enough in the face of future threats.

It is a question that has not received much attention in most member states. There has not been a real debate about the desirability of EU crisis management. What threats should the Union address? What should the EU do and what should be left to the member states? And if an agreement can be reached, how should it be implemented? There are few widely agreed-on answers to these questions.

In this final chapter, we offer arguments and evidence that can help kick off such a debate. However, before we can discuss ways to move forward, we must look back to see how the EU has developed the capacities it currently has.

Patterns of institutional development

One of our more intriguing findings is the unidirectional development of the various EU policy fields dealing with the different crisis types: capacity building rolls inexorably forward. While each policy area discussed in this book has seen long institutional lulls in which nothing seemed to happen, such spells were often followed by bursts of energy that gave rise to new policies and new developments.

The steady rise in EU crisis management capacity, and actual cooperation in crisis management in the EU, poses an intriguing puzzle: how did the EU manage to expand in areas that typically fall within a nation's sovereignty – and did so in times that were decidedly unfriendly to expansionary initiatives?

This "one-way trajectory" is certainly not the exclusive outcome of farsighted leadership or strategic blueprints. Our book points to a combination of "mission creep" and what policy scholars call "incrementalism" (Lindblom 1959), and what in the EU literature is called "functional spillover" (Lindberg and Scheingold 1970; Niemann 2006). This is true both for the goals and for the way these shifting goals have been addressed through capacity building.

Looking across the three crisis domains, we note remarkably similar patterns of goal development: they gradually widened irrespective of and seemingly unguided by original policy intentions.

In the area of external crisis management, the zone of operations gradually increased from the Balkans to the world at large; the strategic threats first identified in the European Security Strategy steadily broadened. The type of missions expanded from purely military to a variety of civil–military "hybrid" missions. The Civil Protection Mechanism has been progressively "recast" to cover both internal and external crises and to apply to a widening number of objects of protection (e.g. people, property, and cultural heritage) and threats. The Civil Protection Mechanism began as a tool for mutual assistance to assist member states, but it is now assuming properties of organizations found at national levels (the US's Federal Emergency Management Agency, for example), including pre-positioning of resources and nurturing shared risk management policies. The emerging domain of transboundary threat cooperation has been marked by a broadening conception of internal security: from criminal justice and border issues to terrorism and critical infrastructures. The Internal Security Strategy now talks about an EU Security Model built on an all-encompassing notion of "protecting EU citizens" from an assortment of cross-border threats.

In a way this reflects a classic pattern of institutionalization: organizational and policy tools become "infused with value beyond the technical requirements at hand" (Selznick 1957: 17). They may then eventually come to define an EU policy space (Stone Sweet, Fligstein, and Sandholtz 2001), which typically happens through a process of trial and error (cf. Sabel and Zeitlin 2010; Zeitlin 2011).

For all three types of crises, experience and learning were driving factors in the development of cooperation structures and procedures. In the area of external crises, problems occurring during foreign missions led to relatively rapid changes in administrative and decision-making structures, which in turn boosted the EU's ability to conduct increasingly complex missions. A growing track record fed the demand for the EU's rather unique crisis management services.

The prospect of disasters that overwhelm a single member state gave rise to a modest decision in 1987 to allow the Commission to begin a network for national civil protection officials. Over time, the Commission's capacity for organizing joint assistance gradually improved as a result of actual support operations. When this capacity proved useful in organizing a coordinated EU response to foreign disasters, the Mechanism quickly gained visibility on the radar of the EU's CSDP apparatus. The tool, in all its simplicity, is now a potential anchor for both external and internal security capacity building (Lindström 2010).

The area of transboundary crises, in particular, is characterized by "learning on the fly." The flawed response to the ash cloud crisis and the confused approach to the financial crisis have prompted a series of hitherto unthinkable changes that have greatly enhanced the EU's capacity to deal with transboundary threats. Given the number of actors and initiatives we identified in this budding domain, more capacity appears to be in the offing.

Vectors of institutionalization

This book has taken a self-consciously light touch when applying theoretical frameworks to this emerging empirical field. We applied an explanatory framework based on institutionalist assumptions of policy development – but we steered clear of forcing one particular theoretical orientation onto the data. Our main goal is to illuminate a new field of European cooperation for scholars of different theoretical stripes.

The institutional trajectories described in this book by and large followed the model set out in Chapter 1. This model views institutionalization as the resultant of interplay between vague yet ambitious goals, bureaucratic experimentation, lesson drawing, policy revision, and feedback. By this account, a modest amount of legal authorization is given to the EU institutions (usually the Commission) to initiate new programs. Over time and through accumulated experience, EU officials demonstrate the utility of cooperation to national officials (mostly lower-ranked experts) who in turn grow accustomed and even comfortable with the role of "Brussels" (Zürn and Checkel 2005).

When sector-specific activities reach the legal limits of their authority, EU officials, often coordinating with national counterparts, pressure governments to cede additional authorization for cooperation. At the next intergovernmental conference (when treaties are revised), new legal bases may be opened for cooperation (Christiansen and Jørgensen 1998). This provides the EU with enhanced tools, which, in turn, invites requests for more EU action (reinvigorating the cycle).

While this model may seem somewhat teleological, we do not imply that institutionalization can only move in one direction. Institutionalization, especially in its early phases (where many capacities still appear to be), can easily grind to a halt and reverse. Our empirical examination of the three policy areas unearthed a set of dominant vectors that can propel or slow down institutionalization processes.

The unintended consequences of success: spillover as a source of new initiatives

Paradoxically, many of the EU's crisis management initiatives and developments find their roots in earlier EU accomplishments. Building a single market, for instance, erases borders and creates efficiency gains. It also generates new interdependencies, many with positive aspects (such as making states reliant on an interconnected set of energy networks to increase the reliability of supply). This process inevitably generates negative externalities (such as the ability of criminals to easily cross borders) and highlights new regulatory gaps (national security regulations may not apply to transboundary infrastructures).

Spillover takes place when states agree to allow some degree of supranational responsibility for accomplishing a limited task and then discover that satisfying that function has external effects on other interdependent activities. The effects of spillover often trigger new policies – or "flanking measures," to use the Brussels terminology (Bulmer 1998). These new initiatives (and evaluations of existing initiatives) are typically framed as addressing the effects of cooperation externalities (Haas 1968; Niemann 2006), in turn requiring further delegation of supranational responsibility.

The effects of spillover are especially dominant in explaining the institutionalization of transboundary crisis management capacities. In the case of transboundary crises, the steady, if scattered, growth of capacities can be traced back to efforts to "make safe" many of the initiatives associated with building the Schengen zone, the single market, and the Single European Sky.

The CSDP capacities and the external use of civil protection capacities may be viewed as a consequence of EU citizenship introduced by the Maastricht Treaty (1992), which clarifies that EU citizenship is "complementary" to national citizenship. This created protection responsibilities for the EU. The Lisbon Treaty sections on fundamental rights reflect an expanding conception of citizenship, which will likely bring additional spillover.

Only the capacities having to do with the internal use of civil protection do not fit the spillover pattern: natural disasters are not the result of European integration. The familiar rationale for cooperation as "leveling the playing field" does not provide a compelling reason for EU action in this field.

Critical events as drivers of ambition

For spillover to translate into crisis management capacities, more is needed than a problem: we need an *urgent* problem. Most, if not all, initiatives to create EU crisis management capacities find their origin in a complex problem that challenges and confounds member states and EU institutions. It can be an unintended consequence of the EU's integration drive (migration, crime), the sudden emergence of a new type of crisis (mad cows and ash clouds), or the EU's embarrassing shortcomings to accomplish its own ambitions ("this is the hour of Europe").

The finding that spillover needs to be imbued with a sense of urgency fits with conventional wisdom among policy scholars, who have noted that radical change often follows on the heels of a critical event (Kingdon 1984; Baumgartner and Jones 1993; Birkland 1998). After a crisis or disaster, public concern rises and political leaders are under pressure to "do something." This creates an opportunity to introduce changes or initiate reforms that would be unthinkable in normal times (Alink, Boin, and 't Hart 2001).

The EU fits this pattern. The Seveso toxic chemical release and the increasingly severe forest fires in the southern member states helped to build support for a "civil protection mechanism." The Balkan conflicts nurtured agreement on the apparent need to build joint military capacity. Swiftly moving threats such as SARS, the ash cloud, and the financial crisis prompted acceptance of far-reaching initiatives to enhance the EU's transboundary crisis management capacities.

We see another phenomenon described in the literature: the EU tends to engage in what scholars call "single-loop learning" (Argyris and Schön 1978). After a crisis, action is initiated to address the causes of the event and the botched response to it: that crisis will never be allowed to happen again. Such events rarely give rise to improved capacities that can also enhance the response to other types of crisis.

These "focusing events" may spur immediate action, but they do not necessarily have a long-term effect on institutionalization. National leaders can call for enhanced cooperative crisis management in the wake of disaster, but they are not always willing to implement crisis-induced initiatives after the initial shock has subsided (cf. Wagner 2003; Princen and Rhinard 2006; Boin, McConnell, and 't Hart 2008; Patashnik 2008). This explains why we often see a disconnect between rhetoric and reality: broad, sweeping statements that sell well "back home" often lack the specificity to be effectively implemented and lack the long-term political support to ensure initiatives are carried through by lower-level national officials responsible for implementation. This is not a risk-free strategy, as expectations may be created against which national leaders will be assessed during the next crisis.

The constraints of legal competences

Institutionalization is both constrained and facilitated by the breadth of preexisting legal competences for EU action. Legal competences, at their most basic, reflect decisions by member states to allow the EU institutions to facilitate cooperation and then to delegate autonomous authority to those institutions (Meunier and Nicolaidis 1999). Cooperation on different types of crises is subject to different legal bases (Antoniadis, Schutze, and Spaventa 2011).[3] If a legal basis is weak, new initiatives are likely to be cautious and modest. If a legal basis is strong, new initiatives can be more daring and adventurous (Craig and de Burca 2007).

The effect of varying degrees of legal competences becomes apparent when comparing human and animal disease control. The latter has a strong legal basis underpinning cooperation, which provides the Commission with significant powers to create, and to enforce, standards and common provisions. In the area of human diseases, however, the Commission's hands are tied. As a result, the Commission must "work around the edges" of its existing legal competences. This can create problems of bureaucratic turf when it is not clear whether it is an animal or a human disease.

The fact that EU competences in the area of development aid have a strong legal basis allows the Commission to push the boundaries of cooperation quite far in comparison to, say, humanitarian relief where legal competences are not as solid. In the area of civil protection, the Commission's authority to facilitate cooperation was heavily constrained through the 1990s since civil protection did not have its own legal basis (another legal basis referring to internal market-related initiatives was used). The "Mechanism" decision in 2001 changed matters and generated an expansion of civil protection initiatives. The inclusion of a proprietary legal basis for civil protection in the Lisbon Treaty (Art. 196) will likely have a bolstering effect.

When "new" crises emerge (such as the volcanic ash cloud or the financial crash), a lack or absence of legal competences is felt immediately, prompting a search for ad hoc arrangements that carry the potential of unintended consequences and mission creep. Even without strong legal bases, the Commission still has room to maneuver: one reason there has been a proliferation of rapid alert systems and "crisis centers" within the Commission is that these are seen as administrative tools rather than policy initiatives.

The importance of organizational leadership and entrepreneurship

Some political scientists are wont to portray international organizations as weak actors that are at the mercy of their members to accomplish anything. International relations scholars, in particular, employ a variety of theories exemplifying this assumption (Waltz 1979, 2000;

Gilpin 1981; Grieco 1996; Pease 2012). They see international organizations as the rational result of their members' wishes (Mearsheimer 1994; Moravcsik 1998). Other scholars recognize the influence international organizations' leadership can have over cooperative policy outcomes (Finnemore and Sikkink 1998; Barnett and Finnemore 1999; Howorth 2010).

Even though our model (laid out in Chapter 1) did not explicitly train our attention on leaders or leadership, our findings point to a clear role for the people working in EU organizations and shaping the policy developments and bureaucratic experiments we described.

Far from the average international secretariat, the Commission is a legally independent institution with the power to initiate policies, hold governments to account, and "breath life" into the treaties (Vahl 1997). In the aftermath of a crisis, the Commission is in a position to present new initiatives and provide "substance" to Council declarations. The Commission seizes on such opportunities to push its own agenda and stake out new policy ground (Christiansen 2006; Rhinard 2010). It provides momentum and, when needed, legal muscle behind the EU's push to enhance crisis management capacities.

The Commission is not the only EU institution with an entrepreneurial approach to policymaking. On the Council side of the EU house, Javier Solana established himself as "Mr. CFSP" and was a driving force in generating member state agreement on CSDP missions. He also led a drive to draft the European Security Strategy in 2003 that won the support of all member states against a backdrop of international discord and discontent following the US invasion of Iraq. Various Council committees, most notably the PSC and the EUMC, have consistently displayed an internal sense of common identity that gives them some autonomy vis-à-vis the member states.

The same is true for EU agencies. In theory they are tightly controlled by the member states, but in practice there is much room for these organizations to develop a distinctive take on the formal aims of the agency. Making use of their autonomy, these agencies have become quite important in shaping how policies are implemented (Groenleer 2013). The European Central Bank – admittedly much more independent than most EU agencies – epitomizes the importance of international organizations: it has arguably become the most important institution in the EU's response to the ongoing financial crisis management.

In short, there are plenty of EU organizations that act as "policy entrepreneurs" in pursuit of change. Although it is too crude to define such interests solely in terms of "more Europe," the Commission certainly measures its performance in terms of common policies and EU-managed programs (Nugent 2001). We found the same spirit in the Council committees.

Policy lock-ins

The importance of policy entrepreneurship is tempered by the effects of so-called policy lock-ins: we refer here to mechanisms in the policy process that make it hard to deviate from established patterns and practices. When an initiative to make things safer meets with perceived success, and a feedback loop confirms the importance and promise of that initiative, employees, politicians, and stakeholders become invested (Pierson 2001; Kuipers 2006). As more and more people come to rely on the new policy or mechanism, it becomes harder and harder to change it (even if better alternatives become available).

Capacities thus tend to emerge slowly from a combination of persistent problems, political declarations, and institutional tinkering (infused with a touch of policy entrepreneurship). Many policy developments across the EU's crisis management landscape can indeed be traced quite a way back. The text of the Solidarity Clause has circulated in various forms since the early 1990s. Most of the ideas that have come to define CSDP capacities (the Petersberg tasks, civil–military missions) have a long policy history and by themselves do not present major breaks ("punctuations") with history. The "headline goals" concept was first used for military capacity building then for civilian capacity development. The original proposal for the European Voluntary Humanitarian Aid Corps dates back to 2003 (introduced under the Greek Presidency).

As we have seen, the EU does not like to reinvent the wheel when it comes to crisis and disaster management. New policy developments often build on earlier ideas or evident successes. The creation and evolution of civil protection cooperation was, for instance, largely modeled after the earlier effort to create a marine pollution framework, and the CCA drew inspiration from NATO (Wendling 2010). The EU's pandemic surveillance and control approaches rely heavily on established WHO practices. In policy areas where agreements among member states have been negotiated, few officials want to propose major and possibly controversial new avenues of change: incremental adjustment becomes the norm.

It is often observed that path dependency leads to stability, if not stasis. Our work affirms that change certainly is possible, just less so in the short run. If we take the capacities as they exist today and compare them with the situation of, say, two decades ago, the difference is overwhelming. Gradual adaptations (which may evoke a picture of tranquil stability) can take a given policy or entity a long way from its (often modest) origins.

When it comes to crisis and disaster management capacities, adaptation is almost guaranteed to take the field down the path of institutionalization. Once such capacities become available, they tend to be used. Modest tools are likely to disappoint, which will make further strengthening look both self-evident and urgently required (few tools that disappoint are formally discarded). All this suggests an inexorable, if slow, drive forward. It may be qualified as "driving with the brakes on," but it is development all the same. If development of crisis management capacities has a sense of inevitability to it, we should ask how to assess the performance of such capacities. We do this next.

From analyzing institutions to normative assessment

Any assessment of the European Union's crisis management capacities should start by reiterating that the EU was never formally intended to become an actor in safety and security questions (White 2011). Given the inherent constraints that beset institutional development in the EU, the rise of crisis management capacity in the European Union is nothing short of spectacular. The broadening horizon of crises addressed at the EU level signals a new and exciting facet of the EU's evolving international security identity. Against a backdrop of relentless euroskepticism, the growth and legitimacy of these crisis capacities offer a clear example of what the EU can do for people at home and abroad.

But while the existing capacity may be more than one could reasonably expect in the light of the constraints that characterize EU policymaking, it may be less than what is needed (see Boin *et al.* 2005). If we compare the available capacities to a list of requirements for effectively managing crises and disasters, and if we consider the types of crises the

EU is likely to face in the future, we must conclude that the EU's crisis management capacity is limited at best. If we combine this with our observation of under-use, it becomes clear that much work remains to be done.

The under-use of limited capacities

The response to a large-scale crisis or disaster typically requires a network of organizations, a set of resources, and some coordinating capacity. This latter capacity – leadership of the response network – is the role most likely to be played by the EU. The role description sounds deceptively simple. As we explained in Chapter 1, the actor in charge of the response network must have the capacity to make sense of an evolving threat; it must be able to initiate a timely and effective response; and it must be able to communicate effectively (Boin *et al.* 2005). In light of these requirements, we conclude that the EU still has some way to go.

The EU's capacities to manage crises differ for each crisis type. The capacities are perhaps most defined when it comes to the support of overwhelmed member states (see Chapter 2). Yet the Civil Protection Mechanism, as it stands today, is unlikely to add much value when an overwhelmed member state calls for help.

The EU can do more, and has accomplished more, in responding to "foreign" crises (Chapter 3). It brings to the table a rather unique set of competences consisting of both military muscle and civilian expertise, which few international organizations can match. The EU has limited its foreign missions in terms of scope and aims, so we do not know for sure the full extent of the EU's external crisis management capability.

The EU possesses a rather broad array of "transboundary capacities," but the results have been underwhelming at best (think of the Iceland ash crisis and the financial crisis). One of its most promising tools – the CCA – has hardly been tried (member states reportedly balk at participating in exercises that are aimed to familiarize them with these arrangements). The member states do not seem ready to enable the EU to take the lead in the face of transboundary threats.

More generally, we encountered four prominent shortcomings:

(1) The EU lacks clearly defined crisis management venues. To be sure, the venues exist and they are being constantly adapted. But they

are either far from transparent, lack an organizational logic that makes sense to the uninitiated, or are very complex.

(2) The EU does not have a strongly developed capacity to collect and analyze data in *real time*. The EU sports quite a few early warning systems, information networks, centers, and crisis rooms – and they are being enhanced. But, on close inspection, they all depend on the willingness of member states and international organizations to provide access to crisis data.

(3) The EU lacks recognized authority to make things happen during a crisis. We have noted that the EU was never intended to become a crisis manager or even a crisis coordinator. But the EU has become one, without the authority to make quick decisions that can guide action in the field. The EU has developed an impressive patchwork of "work arounds" to make up for this deficit, but that is unlikely to suffice in the face of large-scale disasters.

(4) The EU does not use its crisis management capacities in a predictable way. It does not have transparent criteria for the employment of its tools and mechanisms: Which crises and disasters are "deserving" of EU attention? Which capacities will be made available?

This may explain why the EU's crisis tools are not heavily used. Capacities for military crisis management missions are sparingly used, and the desire for military adventures appears to be on the wane (civilian crisis management missions appear to be more popular). We see a similar dynamic in civil protection cooperation, where use of the civil protection mechanism is rarely used for internal disasters. When it is used, EU member states from southern Europe make requests far more often than EU member states from the north, despite the fact the latter have had opportunities to do so.

This is even true for the EU's transboundary crisis management capacities. Calls for improved intra- and inter-institutional cooperation are often accompanied by references to the "transboundary effects" of modern crises (CCA, Argus) or the "cross-border" risks associated with globalized society (such as pandemics, financial crises, and cyber-threats). Yet when these threats materialize, member states find it hard to delegate authority to EU institutions.

So here we encounter a paradox: the EU and the member states have heavily invested, over a long period of time, in the development of crisis management capacities. But development of these capacities, and subsequent use of them, has been slow. What is slowing further development and use of these capacities? We recognize three limiting factors: strategic ambiguity, institutional complexity, and modest funding.

Strategic ambiguity

In studying the EU's capacities, we were struck by the lack of strategic vision that can facilitate thorough discussion of aims and means. For each crisis type studied in this book, goals and strategies are either understated, ambiguous, or do not correspond with available capacities.

The development of CSDP (Common Security and Defence Policy) was never guided by a central philosophy. The ESS is a well-written document, but strategy was determined by missions on the ground and the adaptations made in the face of difficulties encountered. Missions have deployed far beyond the geographical territory for which they were originally intended and they have not been deployed to seemingly obvious hot spots.[4] The same holds for internal crisis management, which remains ill defined in the Lisbon Treaty. The concept of transboundary crisis does not even exist in official EU parlance. A vision exists for civil protection, but it does not correspond with the available resources.

While the EU has been working to produce strategy documents in these three crisis arenas, the lack of vision is most palpable when it comes to a holistic perspective on the EU as crisis manager. Such a holistic perspective will be necessary as the divide between "foreign" and "internal" crises becomes increasingly untenable (Eriksson and Rhinard 2009). This blurring of the internal–external divide is noted in many official documents that the Council and Commission have published in recent years.

A key question for the future is if and how the EU will integrate its capacities to create generic crisis management capacity that can be used to address all types of crisis. For instance, we can easily see how CSDP capacities could be "turned inward" to be made useful in the management of transboundary crises. But while the various policy documents pay lip service to the necessity of bridging the internal–external divide, how that should translate into a reorganization and renewal of capacities remains elusive.

In politically sensitive situations, it can be effective to work under the cloak of strategic ambiguity. But in the absence of a broader debate on the direction and use of EU crisis management capacities, an expectations–capability gap may develop where people expect or demand too much from the EU in times of acute adversity (cf. Hill 1993).

This gap is widened by political rhetoric that does not match developments "on the ground." Council conclusions typically deliver sweeping statements that promise the moon, and actual achievements are routinely exaggerated. For instance, several of the EU's military missions were too ambitious and underestimated the complexity of the task. Keen to score an early success, the EU found itself learning lessons the hard way. The creation of Battle Groups has been trumpeted as a success, but these newly developed capacities are not used in practice (even when the situation calls for their use – Haiti and Libya serve as instructive examples). In the area of civil protection, policy initiatives are frequently prefaced by sweeping statements regarding solidarity between member states and the importance of a "European approach" to disaster management, which belies the actual use of these capacities. When large crises occur, few member states call the Monitoring and Information Centre into action.

To some extent, the systematic overselling of EU initiatives is endemic to the process of European integration, not least because garnering support for such initiatives requires a broadly crafted message (Rhinard 2010). Formulating sweeping ambitions comes with few direct costs. But when the gap between promises and performance becomes too large, it may begin to undermine the credibility and legitimacy of crisis management capacities.

Institutional complexity

The vagaries imposed by the EU's institutional complexity form a second factor limiting the development of crisis management capacities. The EU's crisis management operations are characterized by intense difficulties of coordination within and between the EU institutions as well as between the EU and member state levels. In response to these complexities, the EU continues to develop and design new coordinative arrangements (often quite effective), which, in turn, make the institutional fabric increasingly dense.

The institutional web of directorates, agencies, tools, and tasks can be quite opaque. The complexity of the institutional network that governs CSDP missions is reflected in the chains of command, which tend to have a rather ad hoc character. The case of cooperation on transboundary crises reveals a similarly complex (and disjointed) institutional setting of units, organizations, networks, and instruments.[5]

Institutional complexity is not inherently problematic. The decision-making infrastructure for CSDP missions looks complex on paper but seems to work, with officials settling into a workable routine. Somehow, they get the job done. In fact, institutional complexity may serve the purpose of ensuring efficiency: different political and administrative actors, representing different professional and organizational interests, can be included in the decision-making processes.

But institutional complexities tend to translate into institutional redundancies, which, in turn, can undermine legitimacy. For external crises, the Council developed its Civilian Response Teams, while the Commission created its own Advanced Planning Teams, which have overlapping tasks. In the area of early warning and crisis communication, both the Commission and the Council have developed similar competences (or have ambitions to develop similar competences), reflected in the various "crisis rooms" that can be found in EU institutions. The Lisbon Treaty has given rise to reforms addressing these redundancies, but the EU's institutional mechanisms have a tendency to produce new ones. Such redundancies can easily become subject to politicization in a climate of financial austerity.

Modest funding

A third factor is found in the area of financing. The material resources to carry out the EU's ambitious goals remain slim. The CSDP budget for the Financial Framework for 2007–13 amounted to a mere 0.2% of the total EU budget.[6] For internal civil protection cooperation, there has been a gradual, but still small, accumulation of funds as part of the "financial mechanism" agreed by member states in 2007. This mechanism offers roughly €190 million over five years for training purposes and the hiring of airlift capacity during a disaster. For cooperation on transboundary crises, such as pandemics or infrastructure failure, EU-level financing is less than what national governments spend on such issues. Transboundary cooperation related to migration crises has seen a gradual increase in funding, through such initiatives as Frontex and the Solidarity Fund, but those amounts are also quite low if we compare them to national spending. Humanitarian aid in response to external disasters is the exception. There, funding available represents a wider share of total EU spending.

Epic battles over financing European cooperation punctuate the history of the EU, with still visible scars reflected in the stubborn opinions of the "1%" club of nations refusing to allow EU spending to climb too far above 1% of total EU GDP (Wall 2008). These member states prefer the EU's approach to governance that tilts heavily toward such low-cost activities as regulation, standard-setting, and norm transfer (Majone 1996; Manners 2002). In the light of the financial problems besetting most member states, low levels of financing are likely to shape major crisis management initiatives at the European level in the foreseeable future.

Some hard questions before moving ahead

The gradual growth of EU crisis management capacities has evoked few questions and very little debate in the member states (the EU's financial crisis management being the obvious exception). Consequently, there are few agreed-on questions with regard to the design and functioning of the institutional landscape of European safety and security.

The EU is in the middle of a "change window" in the wake of the adoption of the Lisbon Treaty: actors are figuring out what the new treaty means in terms of how the EU operates as a whole (procedural changes), who leads policy processes (leadership changes), and what ideas will guide future growth (conceptual changes). As it is much harder to affect policy direction after institutionalization sets in (due to the earlier mentioned policy lock-ins), now is the time for such a debate. Three developments, in particular, prompt important and potentially divisive questions.

The first question pertains to the Solidarity Clause. The Clause demands that "the Union and its Member States shall act jointly in a spirit of solidarity if a Member State is the target of a terrorist attack or the victim of a natural or man-made disaster." The Union will mobilize "all the instruments at its disposal, including the military resources made available by Member States" to combat a long list of threats: terrorism, man-made disasters, and natural disasters.

The Lisbon Treaty's Solidarity Clause thus seems to proffer a blanket application of solidarity. But it is not self-evident how solidarity should be interpreted. So it is not clear if member states must assist one another in the event of a crisis (and what type of crisis) or to what extent member states can count on assistance. An addendum to the Solidarity Clause makes clear that the type of assistance provided will be decided at the discretion of each member state. Whether solidarity can be "constitutionalized," and how exactly solidarity will be demonstrated in practice, will have implications for the EU's civil protection and internal security capacities (Myrdal and Rhinard 2010).

A second question concerns the EU's broad approach to external crisis management. A new sense of security thinking – eloquently put forward in the EU's European Security Strategy – has come to dominate many of the EU's traditional institutions and areas of security cooperation. It is in this way that the Union asserts its presence in the "global risk society." The EU is defining a new role for itself on the international stage (Ekengren and Simons 2011). Combining – indeed, integrating – an internal focus on facilitating societal security and an external role as "good governance supplier," the EU is transcending the traditional distinction between internal and external security. This is a unique model, which borrows elements from one type of crisis management capacity to build another.

In adopting this new model, the Union is engaging in a major policy gamble that these modern conceptions of security will translate into a more secure Europe. Will an EU mission to train police forces in Afghanistan make European society safer? Will an EU mission to improve the quality of prisons in Georgia make European citizens more secure? Will the absence of a truly integrated EU army hurt security of the member states? Will the EU's presence in faraway hot spots invite terrorist attacks back home?

As the EU continues to enhance its capacity and employs it across the world, we can expect demand for it to rise – both at home and abroad. This will require criteria for when the EU should, or should not, become involved. As of now, the EU lacks strategic clarity on when to trigger cooperation on the three types of crises studied in this book.

Finally, the rise of EU crisis management capacities evokes questions of democratic legitimacy and accountability. It remains unclear how accountability for the EU's crisis management actions (and non-actions) is arranged. In a formal sense, there are intricate arrangements for the various policy domains: the Council directs foreign missions and the Commission is responsible for civil protection and transboundary crisis management. But as the lines are blurring between external and internal security, and institutional arrangements are being adapted to deal with these changes (the "double-hatting" of Catherine Ashton comes to mind), it is not clear who is responsible for what anymore. That makes accountability particularly hard to organize.

If we can take one lesson from the recent crisis experiences of member states (and other countries), it must be that contemporary crises are marked by growing politicization (Boin, 't Hart, and McConnell 2009). The growing role, and the accompanying rise in visibility, of the EU in transboundary crisis management operations is likely to enhance expectations, attract more media attention, and invite scrutiny from politicians (in the European Parliament and in the member states).[7]

Charting a way forward

Much has been accomplished in a relatively short span of time. It is true that many of the EU's instruments – especially in the transboundary crisis arena – may be downplayed as toothless and the relations between EU institutions and member states certainly remain fragile. But the EU's safety and security apparatus is still under construction. We have observed a sense of excitement when talking to our respondents in Brussels. They confirm our finding that the EU continues to enhance capacities, inching forward by launching initiatives and making a difference, however modest it may seem.

How can the EU best move forward? From a long-term perspective, the EU would be best advised to keep doing what it has done so effectively over the past decades: nurture institutionalization processes by gradually stretching mandates and shaping these processes "on the ground."

The question is whether this will work in the short run. New and potentially destructive crises that outstrip national capacities are likely to emerge, both at home and abroad. Whether we are talking about immigration flows or sovereign debt crises, it is becoming painfully clear that member states cannot deal with these problems alone. Collaboration between member states on crises and disasters is a collective action problem that may well require new ways of thinking (Lagadec 2009; Rhinard 2009).

Now, then, is the time, especially in an era of constitutional change and procedural reform, to consider what needs to be done. Here, at the end of this book, we formulate the most pressing points of action that we think should be taken up both at the national and the European levels.

Define objectives

The EU has long enhanced its crisis management capacities without openly debating or explicitly agreeing on the objectives of crisis management. There is little conceptual guidance as to what an "overwhelmed" member state looks like. In the area of transboundary crises, there is little discussion of what common actions might be taken and by whom. For all crises, the idea of "solidarity" is left largely undefined and unspecified. Can a financially strapped country call on the EU under the solidarity banner?

While this "strategy of ambiguity" – if we can speak of a strategy – has certainly proved a fruitful one, the next stage is best entered with a shared idea of what objectives the EU seeks to accomplish when confronted with critical events in faraway hot zones or closer at home. We propose a basic set of questions that should spur debate (Boin and Ekengren 2009). Answers to these questions can help to steer discussion and gauge the level of commitment among member states.

The first question asks *what* the EU should safeguard through its various crisis management capacities. Outside its borders, the EU is working toward several aims including the protection of human rights, the stabilization of international order, the safety of EU citizens, and the rebuilding of disaster-stricken regions. The ESS makes a valiant effort to argue a connection between these aims and the well-being of EU citizens, but that relation does not hold up to even casual scrutiny.

Both the Civil Protection Mechanism and the EU's transboundary crisis management capacity are aimed at assisting member states in their duties and efforts to protect citizens and the critical infrastructures that make life as we know it possible. The EU's relatively recent Internal Security Strategy (2010) set out to achieve some conceptual clarity in this area, most notably in the articulation of a "European Security Model," but that effort, due to a confused and conflictual drafting process, did not achieve a useful level of consistency and clarity.

The second question asks *how* the Union should achieve its crisis management objectives. Common capacity to a large extent is built in informal, transgovernmental networks. The Union has chosen to pool military and civilian means with the help of intergovernmental, diplomatic coordination (Howorth [2007] calls it "coordigration"). The member states delegate limited control and provide "coordinative authority" to the EU in times of crises but do not cede sovereignty in the irreversible, formal way that we are used to in other EU areas of cooperation. It is not clear whether this is the most effective way to build joint crisis management capacity in the light of the challenges posed by future crises.

A third question asks: *where* are we headed? Some think that the EU's CSDP is "destined to remain firmly intergovernmental" (Hyde-Price 2008: 164). Others appear slightly more optimistic that these capacities will slowly become "communitarized" (Duke and Ojanen 2006; Kirchner and Sperling 2008).

We believe that the long-term development of EU crisis management capacity would benefit from a contemplative discussion on heuristic "end-points" in relation to which European integration can be understood, explained, and measured (Ekengren 2005, 2008). Does effective crisis management at the supranational level require some form of federalization? Is there a normative end-point for crisis management cooperation or security integration? What would an integrated Europe look like in an era of international and transboundary crisis, security, and defense?

The Solidarity Clause might provide impetus for a discussion on these questions. If we take an expansive view of the clause, it could mean that member states should be able to rely on each other to the point that they fully take the collective European capacity into consideration when building their own national capability (Ekengren 2005, 2008). It could conceivably give rise to a joint crisis management capacity that integrates 28 crisis management systems for transboundary and global crisis management. This would require levels of mutual trust reminiscent of the 1950s when a select group of nation states created the European Community (Deutsch *et al.* 1957; Adler and Barnett 1998).

Define crisis competences

A recurring theme in this book is the confusion, or disagreement, with regard to crisis competences: who is authorized to do what exactly? A clear division of labor within the EU has not been agreed on, even if the Lisbon Treaty and subsequent reforms attempted to clarify matters. The response to the 2010 Haiti earthquake demonstrated that the EU has yet more work to do.

This book has also noted (and quite often) the existence of a national–supranational divide. A clear division of responsibility between the national and supranational levels would go a long way toward clarifying roles in the "high stakes" game of crisis management.

To some EU insiders, this suggestion may seem politically naïve. After all, the European project survives largely because the tough questions, including the end-point of integration and the dividing line between the national and supranational, have been left deliberately vague. As the EU forges new, and riskier, policy responsibilities, constructive ambiguity may no longer suffice.

Any discussion about joint crisis management must be founded on an acceptance that European nations will retain their own perspectives on security and are unlikely to cede powers in this sensitive domain. A growing crisis role for the EU will prompt hard questions for member states: Will the EU's newfound enthusiasm for crisis management impinge on national sovereignty? Will member states be forced to participate in joint response operations? Who will pay for these expanding capacities?

This does not mean further development of joint capacities will require a treaty change or can only happen when firmly kept "off the radar." Member state concerns keep EU ambitions in check, ensuring that only small steps that gain legitimacy can be institutionalized. Alignment between national conceptions of security and emerging EU concepts are needed, but they will take time. This is typically accomplished by building consensus amongst experts on the most "logical" responsibilities to be retained at the national level, shared between levels, or delegated to the supranational level. This is a similar kind of procedure used by US authorities in drawing up the "National Response Framework" document, which outlines roles, resources, and responsibilities for counties, states, and the federal government (US NRF 2008).

Member states, of course, will have to do their part. At the national level, there are multiple policy obstacles to effective cooperation with the EU. For instance, it is not always clear which ministries and agencies are responsible for national relationships with the EU level. There is no "transboundary" authority at the national level responsible for European contacts (although some countries are building such within prime ministers' offices). In fact, there is scant awareness of the EU's capacities. A holistic approach to crises and disasters requires, at the very least, that relations between Brussels and member state authorities do not have to be discovered during a crisis.

We have noted how the EU is working on preparedness guidelines to be fulfilled through "national action plans" (Ekengren 2006). Those plans could be evaluated on a yearly basis together with officials from the EU institutions. In this way EU cooperation is able to respect national diversity in this field at the same time as it promotes the longer-term development of common outlooks, resources, and trust.[8]

Strengthen existing institutions

The emergence of structures and mechanisms that facilitate joint crisis management at the EU level is still a recent phenomenon. As the level of institutionalization is relatively low (most of the crisis management structures are in a "nascent" or "ascending" stage of institutionalization), there will be ample room to "redesign" structures in the coming years. The second Barroso Commission, acting on the changes foreseen in the Lisbon Treaty, has done exactly that. Yet still more change may be required as new challenges appear on the horizon – the result of financial weaknesses, climate change, and geopolitical and technological changes.

One step forward would be to strengthen DG Home's remit to deal with internal crises and disasters. In certain respects, it made sense to shift responsibilities for civil protection cooperation to the Commissioner for International Cooperation, Humanitarian Aid, and Crisis Response. The Civil Protection Mechanism had become primarily a mechanism for external disaster response, and greater synergies may be found by including it in the humanitarian aid portfolio. This shift, however, opens space for new consideration about internal disaster response and crisis management, a role nominally shared by the Commissioner mentioned above and the Commissioner for Home Affairs. Greater attention should be given to the parameters of the latter's role so substantial administrative capacity can be devoted to transboundary crisis management.

An additional strategy would amount to further strengthening the EU's sense-making capacities. The EEAS has consolidated several intelligence centers, but many early warning systems and crisis rooms in Brussels and beyond remain only loosely connected in a rather complex network. The EU's crisis management capacities would be best served by a few centers that can create situational awareness in real time during a crisis.

More broadly, it would be advisable to connect the EU's crisis and disaster management institutions more tightly with "sister" institutions in the international arena. The EU has built excellent relations with international NGOs that help deliver EU disaster aid in foreign regions. The EU has remarkably cool relations with other relevant actors (NATO being the most obvious one). In addition, the EU institutions are not always taken seriously by international counterparts (the US security and crisis community prefers to work directly with member states). This external dimension to institution building should be considered seriously in order to enhance the effectiveness of EU efforts.

Formulate a new Headline Goal for EU crisis management

If the EU is to move beyond the tradition of "mission creep," it may wish to consider a simple and time-honored approach: think through what is needed to reach defined goals, assess what is available, and devise a plan to address the shortcomings. Such a "policy mapping" exercise would help EU institutions to see the full landscape of capacities, including those that overlap, those that offer useful intersections, and those that do not yet exist. The EU has successfully used this instrument for the development of its CSDP capacities and should do so again to develop *integrated* capacities.[9]

Across the institutions, focus should be placed on "dual use" capacities: those structures, tools, and instruments which apply to all three types of crises under investigation here (thus, we are perhaps better off discussing "triple use" capacities).

One clear challenge concerns the decision-making structures: can the planning, preparation, and mobilization structures in the Council Secretariat (normally used only for CSDP) be employed for humanitarian missions or civil protection missions? Another challenge links the actual mechanisms used to share resources and deploy teams. How can the Civil Protection Mechanism serve the civilian missions of CSDP (and vice versa)? It would be worthwhile to consider how information and intelligence resources, typically drawn together within the Council's Situation Centre (now the EEAS Intelligence Centre), might be shared between institutions and policy sectors in order to detect spillover effects when a crisis strikes.

For such synergies to work, the EU would have to increase the number of joint exercises and training programs for institutional administrators. This should extend to joint training between national and supranational officials on the evolving EU role in crisis management and obstacles to effective cooperation. These efforts would help to raise awareness and grease the wheels of interaction during crises.

Celebrate success, sell accomplishments

The EU's emergence as an international crisis manager is a stunning accomplishment in the light of the political and institutional complexities marking any change in the EU. It is also an unknown success story. In Brussels and member state capitals, few are aware of what the EU has available when it comes to the joint management of crises and disasters. While this may have helped early institutionalization efforts, it may well hinder future efforts.

We have come away impressed with the capacities that the EU has developed. To drive that point home, some type of cost–benefit analysis might be helpful. We know that member states (and other nations) are left with huge bills after a crisis or disaster strikes (think of Hurricane Katrina or the ash cloud crisis). If a transnational response could bring down the costs of disaster, would it not be worth investing in such capacities?

In writing this book, we have become convinced that the EU should do more in this domain. The threats to the people of Europe are diversifying and increasing, posing challenges that member states cannot handle. The member states need shared capacity to manage crises – especially the transboundary species – and the EU has demonstrated that it can facilitate such coordination. It should be an easy sell: a joint response to a mutual threat is the true embodiment of European solidarity and a cost-effective way to demonstrate what the EU can do for its citizens.

Notes

1 Speech in the European Parliament at the Joint Debate on Foreign and Security Policy, March 10, 2010.

2 The evolution of EU crisis and disaster management capacities contradicts the pessimism on the part of some academics (see, for example, Etzioni 2007) with regard to the possibilities of further integration in the EU.

3 A legal basis refers mainly to primary law, codified in the treaties, that provides for some degree of policymaking at the European level (legal bases may also be provided through individual Council decisions, but these may not exceed the terms of the treaties). For external crises related to CSDP, legal restraints are less influential since member states firmly control policymaking.

4 As Jean-Marie Guehenno (2009: 8) observed, the EU has not defined what makes peace operations effective. What is absent, the former UN Under-Secretary-General for Peacekeeping Operations observes, is a "single unifying political vision."

5 In some cases, especially where EU legal competences are weak (such as cyber security), institutions are not particularly complex.

6 One point of caution to note when discussing CSDP budgets is that there are separate funding mechanisms for civilian and military CSDP missions. In the former case, budgets are determined by the EC's normal budgetary procedures. This is not the case for military missions, as these are predominantly funded by the member states themselves (Missiroli 2003; Scannel 2004; Keukeleire and MacNaughtan 2008).

7 The European Parliament has shown increasing interest in EU crisis and disaster management. We should also note that the CSDP budget for foreign missions requires the approval of the EP.

8 This method for national convergence is used in other EU policy areas and is viewed by some as an indicator of an emerging type of EU governance (Sabel and Zeitlin 2010).

9 We thank Gustav Lindström for this suggestion. See Lindström (2010).

References

Adler, E. and Barnett, S. (1998). *Security Communities*. Cambridge: Cambridge University Press.

Åhman, T. and Nilsson, C. (2009). The Community Mechanism for Civil Protection and the European Union Solidarity Fund. In S. Olsson, ed., *Crisis Management in the European Union: Cooperation in the Face of Emergencies*. New York: Springer, pp. 83–107.

Akman, P. and Kassim, H. (2009). Myths and myth-making in the European Union: the institutionalization and interpretation of EU competition policy. *Journal of Common Market Studies*, 48(1), 111–32.

Alemanno, A. (ed.) (2011). *Governing Disasters: The Challenges of Emergency Risk Regulation*. Cheltenham, UK: Edward Elgar.

Alink, F. (2006). Crisis als Kans? Over de Relatie tussen Crises en Hervormingen in het Vreemdelingenbeleid van Nederland en Duitsland. Dissertation, Utrecht School of Governance, Utrecht University.

Alink, F., Boin, A., and 't Hart, P. (2001). Institutional crises and reforms in policy sectors: the case of asylum policy in Europe. *Journal of European Public Policy*, 8(2), 286–306.

Allen, D. (2009). The Solana era: the development of the EU's CFSP and ESDP during the period of office of the first High Representative and Secretary General of the Council 1999–2009. Paper presented at the European Union Studies Association, Los Angeles.

Allen, D. and Smith, M. (2000). Relations with the wider world. *Journal of Common Market Studies Annual Review*, 38, 101–20.

Allen, D. and Smith, M. (2004). Relations with the wider world. *Journal of Common Market Studies Annual Review*, 42, 95–112.

Allen, D. and Smith, M. (2005). Relations with the wider world. *Journal of Common Market Studies Annual Review*, 43, 109–26.

Anderson, J. (Ed.) (2002). *Transnational Democracy*. London: Routledge.

Ansell, C. and Vogel, D. (eds.) (2006). *What's the Beef: The Contested Governance of European Food Safety*. Cambridge, MA: MIT Press.

Ansell, C., Boin, A., and Keller, A. (2010). Managing transboundary crises: identifying building blocks of an effective response system. *Journal of Contingencies and Crisis Management*, 18(4), 205–17.

Antoniadis, A. (2011). Debt crisis as a global emergency: the European economic constitution and other Greek fables. In A. Antoniadis, R. Schutze, and E. Spaventa, eds., *The European Union and Global Emergencies: A Law and Policy Analysis*. Oxford: Hart Publishing, pp. 167–91.

Antoniadis, A., Schutze, R., and Spaventa, E. (eds.) (2011). *The European Union and Global Emergencies: A Law and Policy Analysis*. Oxford: Hart Publishing.

Argomaniz, J. (2009). Post-9/11 institutionalisation of European Union counterterrorism: emergence, acceleration and inertia. *European Security*, 18(2), 151–72.

Argomaniz, J. (2011). *The EU and Counter-terrorism*. London: Routledge.

Argyris, C. and Schön, D. A. (1978). *Organizational Learning: A Theory of Action Perspective*. London: Addison-Wesley.

Auerswald, P. E., Branscomb, L. M., LaPorte, T. M., and Michel-Kerjan, E. O. (eds.) (2006). *Seeds of Disaster, Roots of Response: How Private Action Can Reduce Public Vulnerability*. New York: Cambridge University Press.

Bailes, A. (2005). *The European Security Strategy: An Evolutionary History*. SIPRI Policy Paper No. 10. Stockholm: SIPRI.

Balch, A. and Geddes, A. (2011). Opportunity from crisis? Organisational responses to human trafficking in the UK. *The British Journal of Politics & International Relations*, 13(1), 26–41.

Baldwin, R., Gros, D., and Laeven, L. (2010). *Completing the Eurozone Rescue: What More Needs to Be Done?* Brussels: VoxEU Publications.

Barnett, M. N. and Finnemore, M. (1999). The politics, power, and pathologies of international organizations. *International Organization*, 53(4), 669–732.

Barnier, M. (2006). *For a European Civil Protection Force: Europe Aid*. Report commissioned by the Austrian Presidency and Commission President José Manuel Barroso.

Barros-Garcia, X. (2007). *Effective Multilateralism and the EU as a Military Power: The Worldview of Javier Solana*. EUI Working Papers, RSCAS 2007/08. Badia Fiesolana: European University Institute.

Barutciski, M. (1994). EU states and the refugee crisis in the former Yugoslavia. *Refuge*, 14(3), 32–5.

Baumgartner, F. R. and Jones, B. (1993). *Agendas and Instability in American Politics*. University of Chicago Press.

Baumgartner, F. R. and Jones, B. D. (2002). Positive and negative feedback in politics. In F. R. Baumgartner and B. D. Jones, eds., *Policy Dynamics*. University of Chicago Press, pp. 3–28.

Beck, U. (1992). *Risk Society*. London: Sage.

Beck, U. (2005). War is peace: on post-national war. *Security Dialogue*, 36(1), 5–26.

Beck, U. (2008). Reframing power in the globalized world. *Organization Studies*, 29(5), 793–804.

Begg, I. (2010). Economic and social governance in the making: EU economic governance in flux. *Journal of European Integration*, 32(1), 1–16.

Biscop, S. (2004). *The European Security Strategy: Implementing a Distinctive Approach to Security*. Paper No. 82. Brussels: Royal Institute for International Relations.

Biscop, S. (ed.) (2005). *E Pluribus Unum? Military Integration in the European Union*. Egmont Paper 7. Brussels: Royal Institute for International Relations.

Birkland, T. A. (1998). Focusing events, mobilization, and agenda setting. *Journal of Public Policy*, 18(1), 53–74.

Boin, A. (2001). *Crafting Public Institutions: Leadership in Two Prison Systems*. Boulder, CO: Lynne Rienner.

Boin, A. (2005). From crisis to disaster: towards an integrative perspective. In R. W. Perry and E. W. Quarantelli, eds., *What Is a Disaster? New Answers to Old Questions*. Philadelphia, PA: Xlibris, pp. 153–72.

Boin, R. A. and Christensen, T. (2008). The development of public institutions: reconsidering the role of leadership. *Administration & Society*, 40(3), 271–97.

Boin, R. A. and Ekengren, M. (2009). Preparing for the world risk society: towards a new security paradigm for the European Union. *Journal of Contingencies and Crisis Management*, 17(4), 285–94.

Boin, R. A. and Rhinard, M. (2008). Managing transboundary crises: what role for the European Union? *International Studies Review*, 10(1), 1–26.

Boin, A. and 't Hart, P. (2000). Institutional crises and reforms in policy sectors. In H. Wagenaar, ed., *Government and Institutions: Effect, Changes and Normative Foundations*. Dordrecht: Kluwer Academic Publishers, pp. 9–31.

Boin, A. and 't Hart, P. (2010). Organising for effective emergency management: lessons from research. *Australian Journal of Public Administration*, 69(4), 357–71.

Boin, A., Ekengren, M., and Rhinard, M. (2006). Protecting the Union: analysing an emerging policy space. *Journal of European Integration*, 28(5), 405–21.

Boin, A., Kelle, P., and Whybark (eds.) (2010). Improving disaster supply chain management: key supply chain factors for humanitarian relief. Special issue of the *International Journal of Production Economics*, 126(1), 1–143.

Boin, R. A., McConnell, A., and 't Hart, P. (eds.) (2008). *Governing after Crisis: The Politics of Investigation, Accountability and Policy Change*. Cambridge University Press.

Boin, A., 't Hart, P., and McConnell, A. (2009). Crisis exploitation: political and policy impacts of framing contests. *Journal of European Public Policy*, 16(1), 81–106.

Boin, A., 't Hart, P., Stern, E., and Sundelius, B. (2005). *The Politics of Crisis Management: Public Leadership Under Pressure*. New York: Cambridge University Press.

Boin, A., 't Hart, P., and Van Esch, F. (2012). Political leadership in times of crisis: comparing leader responses to financial turbulence. In L. Helms, ed., *Comparative Political Leadership*. Basingstoke: Palgrave Macmillan, pp. 119–41.

Bossong, R. (2008). The action plan on combating terrorism: a flawed instrument of EU Security Governance. *Journal of Common Market Studies*, 46(1), 27–48.

Bossong, R. (2012). *EU Civilian Crisis Management and Organisational Learning*. Economics of Security Working Paper 62. Berlin: Economics of Security.

Bossong, R. and Rhinard, M. (in press). Theorising internal security cooperation in the European Union. *European Security*, 22(2).

Bourdieu, P. (1990). *The Logic of Practice*. Stanford University Press.

Boutruche, S. (2003). Immigration and asylum in the harmonisation policies of the EU: the need for balance. In D. Turton and J. Gonzalez, eds., *Immigration in Europe: Issues, Policies, and Case Studies*. Bilbao: University of Deusto, pp. 75–88.

Brady, H. (2008). *EU Immigration Policy: A–Z*. London: Centre for European Reform Briefing.

Brady, H. (2010). EU JHA co-operation: after Lisbon, reality bites. *Insight*. Brussels: Centre for European Reform, June 24.

Brady, H. (2011). Written evidence. Submitted to the House of Lords Select Committee on the European Union. *Inquiry into the EU Internal Security Strategy* (oral and associated written evidence). London: House of Lords.

Brattberg, E. and Rhinard, M. (2011). Multilevel governance and complex threats: the case of pandemic preparedness in the European Union and the United States. *Global Health Governance*, 5(1), 1–21.

Brauch, H. G., Spring, U. O., Mesjasz, C., Grin, J., Dunay, P., Chadha, N. B., Chourou, B., Kameri-Mbote, P., and Liotta, P. H. (eds.) (2008). *Globalization and Environmental Challenges: Reconceptualizing Security in the 21st Century*. Berlin: Springer.

Brinkley, D. (2006). *The Great Deluge: Hurricane Katrina, New Orleans, and the Mississippi Gulf Coast*. New York: William Morrow.

Brusset, E. and Tiberghien, C. (2002). Trends and risks in EU humanitarian action. In J. Macrae, ed., *The New Humanitarianisms: A Review of Trends in Global Humanitarian Action*. London: Overseas Development Institute, pp. 51–62.

Bulmer, S. J. (1998). New institutionalism and the governance of the Single European Market. *Journal of European Public Policy*, 5(3), 365–86.

Bures, O. (2011). *EU Counterterrorism Policy: A Paper Tiger?* Farnham, UK: Ashgate.

Buzan, B. (1991). New patterns of global security in the twenty-first century. *International Affairs*, 67(3), 431–51.

Buzan, B., Waever, O., and de Wilde, J. (1998). *Security: A New Framework for Analysis*. Boulder, CO: Lynne Rienner.

Calleo, D. P. (2001). *Rethinking Europe's Future*. Princeton University Press.

Carmichael, C. (2002). *Ethnic Cleansing in the Balkans: Nationalism and the Destruction of Tradition*. London: Routledge.

Cavanagh Hodge, C. (2005). *Atlanticism for a New Century: The Rise, Triumph, and Decline of NATO*. Upper Saddle River, NJ: Pearson/Prentice Hall.

Checkel, J. T. (2001). Why comply? Social change and European identity change. *International Organization*, 55(3), 553–88.

Christiansen, T. (2006). The European Commission: the European executive between continuity and change. In J. Richardson, ed., *European Union: Power and Policy-making*. Abingdon, UK: Routledge, pp. 99–120.

Christiansen, T. and Jørgensen, K. E. (1998). Negotiating treaty reform in the European Union: the role of the European Commission. *International Negotiation*, 3(3), 435–52.

Christopher, W. (2001). *Chances of a Lifetime*. New York: Scribner.

Clarke, L. (1999). *Mission Improbable*. University of Chicago Press.

Collett, E. (2010). *Beyond Stockholm: Overcoming the Inconsistencies of Immigration Policy*. European Policy Centre Working Paper No. 32. December.

Comfort, L. K. (1989). The San Salvador earthquake. In U. Rosenthal, M. T. Charles, and P. 't Hart, eds., *Coping with Crisis: The Management of Disasters, Riots and Terrorism*. Springfield, IL: Charles C. Thomas, pp. 323–39.

Cooper, C. and Block, R. (2006). *Disaster Hurricane Katrina and the Failure of Homeland Security*. New York: Times Books.

Coopers, R. (2000). *The Post-Modern State and the World Order*. London: Demos.

Coopers, R. (2003). *The Breaking of Nations: Order and Chaos in the 21st Century*. London: Atlantic books.

Cornish, P. and Edwards, G. (2005). The strategic culture of the European Union: a progress report. *International Affairs*, 81(4), 801–20.

Cottey, A. (2007). *Security in the New Europe*. New York: Palgrave Macmillan.

Coulombier, D. and Ekdahl, K. (2005). H5N1 influenza and the implications for Europe. *British Medical Journal*, 331(7514), 413–14.

Craig, P. and de Burca, G. (2007). *EU Law: Text, Cases and Materials*. Oxford University Press.

Crawley, H. (1993). *The Refugee Crisis Facing Western Europe*. Research Paper 11, University of Sussex Geography Department Brighton, UK.

Crosbie, J. (2007). Countering common threats. *European Voice* (Sept. 27–Oct. 3), 24.

Cross, M. D. (2010). *Cooperation by Committee: The EU Military Committee and the Committee for Civilian Crisis Management*. Paris: European Union Institute for Security Studies.

Cross, M. D. (2011). The European External Action Service and smart power. *EUSA Review*, 24(2), 3–5.

Daalder, I. H. and O'Hanlon, M. E. (2000a). *Winning Ugly: NATO's War to Save Kosovo*. Washington, DC: Brookings Institution.

Daalder, I. H. and O'Hanlon, M. E. (2000b). The United States in the Balkans: there to stay. *The Washington Quarterly*, 23(4), 157–70.

Dalgaard-Nielsen, A. and Hamilton, D. (eds.) (2006). *Transatlantic Homeland Security: Protecting Society in the Age of Catastrophic Terrorism*. New York: Routledge.

Dempsey, J. (2003). Follow my leaders. *Financial Times* (July 12), 47.

Deutsch, K. W., Burrell, S. A, Kann, R. A., Lee, M., Lichterman, M., Lindgren, R. E., Loewenheim, F. L., and Van Wagenen, R. W. (1957).

Political Community and the North Atlantic Area: International Organization in the Light of Historical Experience. Princeton University Press.

de Wijk, R. (2004). European view. In G. Lindstrom and B. Schmitt, eds., *One Year On: Lessons from Iraq*. Chaillot Paper No. 68. Paris: Institute for Security Studies, pp. 41–50.

Diamond, J. (2005). *Collapse: How Societies Choose to Fail or Succeed*. New York: Viking Press.

Dijstelbloem, H. and Meijer, A. (eds.) (2011). *Migration and the New Technological Borders of Europe*. London: Palgrave Macmillan.

Dittrich, M. (2005). *Facing the Global Terrorist Threat: A European Response*. Working Paper No. 14. Brussels: European Policy Centre.

Drennan, L. T. and McConnell, A. (2007). *Risk and Crisis Management in the Public Sector*. London: Routledge.

Dror, Y. (2001). *The Capacity to Govern*. London: Frank Cass.

Dryzek, J. (2006). Transnational democracy in an insecure world. *International Political Science Review*, 27(2), 101–19.

Duke, S. (2000). *The Elusive Quest for European Security*. Basingstoke: Macmillan.

Duke, S. (2002). *The EU and Crisis Management: Development and Prospects*. Maastricht: European Institute of Public Administration.

Duke, S. (2004). The European External Action Service: a diplomatic service in the making? *CFSP Forum*, 2(4) (July), 4–7.

Duke, S. (2005). The linchpin COPS: assessing the workings and institutional relations of the Political and Security Committee. European Institute of Public Administration working paper 2005/W/05.

Duke, S. (2008). The future of EU–NATO relations: a case of mutual irrelevance through competition? *Journal of European Integration*, 30(1), 27–43.

Duke, S. (2010). *The European External Action Service*. Policy Paper 2. Maastricht: European Institute of Public Administration.

Duke, S. and Ojanen, H. (2006). Bridging internal and external security: lessons from the European Security and Defence Policy. *Journal of European Integration*, 28(5), 477–94.

Edwards, G. (2005). The patterns of the EU's global activity. In C. Hill and M. Smith, eds., *International Relations and the European Union*. Oxford University Press, pp. 39–63.

Ekengren, M. (2002). *The Time of European Governance*. Manchester University Press.

Ekengren, M. (2005). From a European security community to a secure European community, *Fornet, CFSP Forum*, 3(1).

Ekengren, M. (2006). New security challenges and the need for new forms of EU cooperation: the solidarity declaration and the open method of co-ordination. *European Security*, 15(1), 89–111.

Ekengren, M. (2007a). EU Civil Protection and the flooding in Austria, the Czech Republic, Germany and Slovakia 2002. In M. Britz, O. Bures, M. Ekengren, H. Moroff, and C. Stålvant, *Challenges and Opportunities for Societal Security in an Enlarged European Union*. Report, ACTA B 39. Stockholm: National Defence College, pp. 27–51.

Ekengren, M. (2007b). The internal–external security challenge for the EU. *Studia Diplomatica*, 55(1), 81–106.

Ekengren, M. (2008). From a European security community to a secure European community: tracing the new security identity of the EU. In H. G. Brauch *et al.*, eds., *Globalisation and Environmental Challenges: Reconceptualising Security in the 21st Century*. New York: Springer, pp. 695–704.

Ekengren, M. and Simons, G. (eds.) (2011). *The Politics of Security Sector Reform: Challenges and Opportunities for the European Union's Global Role*. Farnham, UK: Ashgate.

Ekengren, M., Matzén, N., and Svantesson, M. (2006). *The New Security Role of the European Union: Transnational Crisis Management and the Protection of Union Citizens*. Report, No. B 35 ACTA-series. Stockholm: National Defence College.

Emerson, M. and Gross, E. (eds.) (2007). *Evaluating the EU's Crisis Missions in the Balkans*. Brussels: Centre for European Policy Studies.

Eriksson, J. and Rhinard, M. (2009). The internal–external security nexus: notes on an emerging research agenda. *Cooperation and Conflict*, 44(3), 243–67.

Etzioni, A. (2007). The community deficit. *Journal of Common Market Studies*, 45(1): 23–42.

European Environment Agency (2010). *Mapping the Impacts of Natural Hazards and Technological Accidents in Europe: An Overview of the Last Decade*. European Environment Agency.

Evans-Pritchard, A. (2009). Germany wants to rein in EU financial regulation plans. *Daily Telegraph* (Sept. 22), 21.

Fägersten, B. (2010). *Sharing Secrets: Explaining International Intelligence Cooperation*. Department of Political Science, Lund University.

Fijnaut, C. (2006). Police cooperation and the area of freedom, security and justice. In N. Walker, ed., *Europe's Area of Freedom, Security, and Justice*. Oxford University Press, pp. 253–8.

Financial Crisis Inquiry Commission (2011). *Final Report of the Causes of the Financial and Economic Crisis in the United States*. Washington, DC: US Government Printing Office.

Finnemore, M. and Sikkink, K. (1998). International norm dynamics and political change. *International Organization*, 52(4), 887–917.

Flood, C. (2009). Dimensions of Euroscepticism. *Journal of Common Market Studies*, 47(4), 911–17.

Fritzon, A., Ljungkvist, K., Boin, R. A., and Rhinard, M. (2007). Protecting Europe's critical infrastructures: problems and prospects. *Journal of Contingencies and Crisis Management*, 15(1), 30–41.

Fuchs-Drapier, M. (2011). The European Union's Solidarity Clause in the event of a terrorist attack: towards solidarity or maintaining sovereignty? *Journal of Contingencies and Crisis Management*, 19(4), 184–97.

Gaddis, J. L. (2005). *Strategies of Containment: A Critical Appraisal of American National Security Policy During the Cold War*. New York: Oxford University Press.

Gänzle, S. and Sens, A. (2007). *The Changing Politics of European Security*. Basingstoke: Palgrave Macmillan.

Gebhard, C. (2009). The crisis management and planning directorate: recalibrating ESDP planning and conduct capacities. *CFSP Forum*, 7(4), 8–14.

Giegerich, B. (2008). *European Military Crisis Management: Connecting Ambition and Reality*. Adelphi papers. London: Routledge.

Giegerich, B. (2009). *European Military Crisis Management: Connecting Ambition and Reality*. Adelphi Paper No. 397. London: International Institute for Strategic Studies.

Gilpin, R. (1981) *War and Change in World Politics*. Princeton University Press.

Ginsberg R. H. (2001). *The European Union in International Politics: Baptism by Fire*. Lanham, MD: Rowman & Littlefield.

Goodin, R. E. (ed.) (1996). *The Theory of Institutional Design*. Cambridge University Press.

Gourlay, C. (2006a). Community instruments for civilian crisis management. In A. Nowak, ed., *Civilian Crisis Management the EU Way*. Chaillot Paper No. 90. Paris: Institute for Security Studies, pp. 49–67.

Gourlay, C. (2006b). Civil–civil coordination in EU crisis management. In A. Nowak, ed., *Civilian Crisis Management the EU Way*. Chaillot Paper No. 90. Paris: Institute for Security Studies, pp. 103–22.

Gowan, R. (2005). The Battlegroups: a concept in search of a strategy? In S. Biscop, ed., *E Pluribus Unum: Military Integration in the European Union*. Egmont Papers 7. Brussels: Royal Institute for International Relations, pp. 13–20.

Gowan, R. (2009). ESDP and the United Nations. In G. Grevi, D. Helly, and D. Keohane, eds., *European Security and Defence Policy: The First 10 Years (1999–2009)*. Paris: European Institute for Security Studies, pp. 117–26.

Greco, E., Pirozzi, N., and Silvestri, S. (2010). *EU Crisis Management: Institutions and Capabilities in the Making*. Rome: Istituto Affari Internazionali.

Grevi, G. (2009). ESDP institutions. In G. Grevi, D. Helly, and D. Keohane, eds., *European Security and Defence Policy: The First 10 Years (1999–2009)*. Paris: Institute for Security Studies, pp. 19–68.

Grevi, G. and Keohane, D. (2009). ESDP resources. In G. Grevi, D. Helly, and D. Keohane, eds., *European Security and Defence Policy: The First 10 Years (1999–2009)*. Paris: Institute for Security Studies, pp. 69–114.

Grevi, G., Helly, D., and Keohane, D. (eds.) (2009). *European Security and Defence Policy: The First 10 Years (1999–2009)*. Paris: Institute for Security Studies.

Grieco, J. (1996). State interests and international rule trajectories: a neorealist interpretation of the Maastricht Treaty and European Economic and Monetary Union, *Security Studies*, 5(3), 261–305.

Groenleer, M. L. P. (2013). *European Union Agencies: Autonomy, Institutionalization and Leadership*. Basingstoke: Palgrave Macmillan.

Grönvall, J. (2000). *Managing Crisis in the European Union: The Commission and 'Mad Cow' Disease*. Stockholm: CRISMART/Swedish National Defence College.

Grönvall, J. (2001). Mad cow disease: the role of experts and European crisis management. In U. Rosenthal, R. A. Boin, and L. K. Comfort, eds., *Managing Crises: Threats, Dilemmas, Opportunities*. Springfield, IL: Charles C. Thomas, pp. 155–74.

Gualandri, E. and Grasso, A. G. (2006). Towards a new approach to regulation and supervision in the EU: post-FSAP and comitology. *Munich Personal RePEc Archive* MPRA Paper No. 1780.

Guehenno, J. (2009). Foreword. In D. Korski and R. Gowan, eds., *Can the EU Rebuild Failing States? A Review of Europe's Civilian Capacities*. London: European Council on Foreign Relations, pp. 7n.

Gustenau, G. (2006). The concept of homeland security in the European Union and in Austria: a challenge for the Austrian EU presidency. In B. Esther, ed., *Transforming Homeland Security: U.S. and European Approaches*. Washington: Center for Transatlantic Relations, pp. 59–80.

Haas, E. B. (1968). *The Uniting of Europe: Social, Political and Economic Forces 1950–1957*. Stanford University Press.

Haddad, E. (2008). The external dimension of EU refugee policy: a new approach to asylum? *Government and Opposition*, 43(2), 190–205.

Haine, J. Y. (2004). Idealism and power: the new EU security strategy. *Current History*, 103(671) (March), 107–12.

Hansen, A. S. (2006). *Against all Odds: The Evolution of Planning for ESDP Operations*. Center for International Peace Studies, Norwegian Defence Research Establishment.

Helly, D. (2006). EUJUST Themis in Georgia: an ambitious bet on rule of law. In A. Nowak, ed., *Civilian Crisis Management the EU Way*. Chaillot Paper No. 90. Paris: Institute for Security Studies, pp. 87–102.

Helly, D. (2009) ESDP and the African Union. In G. Grevi, D. Helly, and D. Keohane, eds., *European Security and Defence Policy: The First 10 Years (1999–2009)*. Paris: Institute for Security Studies, pp. 147–58.

Helsloot, I., Boin, A., Comfort, L. K., and Jacobs, B. (eds.) (2012). *Mega-Crises*. Springfield, IL: Charles C. Thomas.

Hendrickson, R., Strand, J., and Raney, K. (2007). Operation *Artemis* and Javier Solana: EU prospects for a stronger common foreign and security policy. *Canadian Military Journal*, 8(1), 35–42.

Hill, C. (1993). The capability–expectations gap, or conceptualizing Europe's international role. *Journal of Common Market Studies*, 31(3), 305–28.

Hodson, D. and Quaglia, L. (2009). Europe and the global financial crisis. *EUSA Review* 22(4), 6–9.

Hogan, M. J. (ed.) (1992). *The End of the Cold War: Its Meaning and Implications*. New York: Cambridge University Press.

Hollis, S. (2009). Patterns of participation: a study on national participation in the EU Mechanism for Civil Protection. Paper prepared for the research program Creating EU Crisis Management Capacity for a Secure European Community.

Homan, K. (2007). Operation Artemis in the Democratic Republic of Congo. In European Commission, ed., *Faster and More United?: The Debate about Europe's Crisis Response Capacity*. Luxembourg: Office for Official Publications of the European Communities, pp. 151–4.

House of Lords (2008). Select Committee on the European Union. In *FRONTEX: the EU External Border Agency*. HL Paper 60. London: HMSO.

House of Lords European Union Committee (2009a). *Sixth Report: Civil Protection and Crisis Management in the European Union. Report with Evidence*. London: House of Lords.

House of Lords European Union Committee (2009b). *The Future of EU Financial Regulation and Supervision*. HL 106-I. London: HMSO.

House of Lords Select Committee on the European Union (2011a). *Inquiry into the EU Internal Security Strategy*. London: House of Lords.

House of Lords Select Committee on the European Union (2011b). *Inquiry into the EU Internal Security Strategy: Oral and Associated Written Evidence*. London: House of Lords.

Howorth, J. (2007). *Security and Defence Policy in the European Union*. London: Routledge.

Howorth, J. (2009) NATO and ESDP: institutional complexities and political realities. *Politique Etrangere*, 4, 95–106.

Howorth, J. (2010). The EU as a global actor: grand strategy for a global grand bargain? *Journal of Common Market Studies*, 48(3), 455–74.

Howorth, J. and Menon, A. (2009). Still not pushing back: why the European Union is not balancing the United States. *The Journal of Conflict Resolution*, 53(5), 727–44.

Huntington, S. P. (1968). *Political Order in Changing Societies*. New Haven, CT: Yale University Press.

Hyde-Price, A. (2004). European security, strategic culture, and the use of force. *European Security*, 13(4), 233–43.

Hyde-Price, A. (2007). *European Security in the Twenty-First Century: The Challenge of Multipolarity*. New York: Routledge.

Hyde-Price, A. (2008). A neurotic "centaur": the limitations of the EU as a strategic actor. In J. Hallenberg and K. Engelbrekt, eds., *European Union and Strategy: An Emerging Actor*. Abingdon, UK: Routledge, pp. 153–66.

Hynek, N. (2011). EU crisis management after the Lisbon Treaty: civil–military coordination and the future of the EU operational headquarters. *European Security*, 20(1), 81–102.

Illies, M. (2009). Irregular immigration policy in the European Union: action at all stages of the irregular migration flow. Real Instituto Elcano Working Paper 38.

Ioannides, I. (2006). EU police mission *Proxima*: testing the "European" approach to building peace. In A. Nowak, ed., *Civilian Crisis Management the EU Way*. Chaillot Paper No. 90. Paris: Institute for Security Studies, pp. 69–86.

Ioannides, I. (2010). EU civilian capabilities and cooperation with the military sector. In E. Greco, N. Pirzozzi, and S. Silvestri, eds., *EU Crisis Management: Institutions and Capabilities in the Making*. Rome: Istituto Affari Internazionali, pp. 29–55.

Joerges, C., Sand, I.-J., and Teubner, G. (eds.) (2004). *Transnational Governance and Constitutionalism: International Studies in the Theory of Private Law*. Oxford: Hart Publishing.

Johns, M. (2011). Social cohesion in the EU: the next challenge for Europe? Paper presented at the annual meeting of the International Studies Association (ISA) Annual Conference "Global Governance: Political Authority in Transition," Montreal, Quebec, Canada.

Jones, S. (2007). *The Rise of European Defence Cooperation*. Cambridge University Press.

Kaplan, L. S. (1999). *The Long Entanglement: NATO's First Fifty Years*. Westport, CT: Praeger.

Kaunert, C. (2010a). *European Internal Security: Towards Supranational Governance in the Area of Freedom, Security and Justice?* Manchester University Press.

Kaunert, C. (2010b). Europol and EU counter-terrorism: international security actorness in the external dimension? *Studies in Conflict and Terrorism*, 33(7), 652–71.

Kaunert, C., Leonard, S., and Pawlak, P. (eds.) (2012). *European Homeland Security: A Strategy in the Making?* London: Routledge.

Kennan, G. (1947). The sources of Soviet conduct. *Foreign Affairs*, 25(4), 169–82.

Keohane, D. (2009). ESDP and NATO. In G. Grevi, D. Helly, and D. Keohane, eds., *European Security and Defence Policy: The First 10 Years (1999–2009)*. Paris: European Union Institute for Security Studies, pp. 127–38.

Keohane, D. (2011). Lessons from EU peace operations. *Journal of International Peacekeeping*, 15, 200–17.

Kettl, D. (2004). *System under Stress: Homeland Security and American Politics*. Washington, DC: CQ Press.

Keukeleire, S. and MacNaughtan, J. (2008). *The Foreign Policy of the European Union*. London: Palgrave Macmillan.

Khol, R. (2006). Civil-military co-ordination in EU crisis management. In A. Nowak, ed., *Civilian Crisis Management the EU Way* (pp. 123–138). Chaillot Paper No. 90. Paris: Institute for Security Studies, pp. 123–38.

Kingdon, J. (1984). *Agendas, Alternatives, and Public Policies*. Boston, MA: Little, Brown.

Kirchner, E. J. (2007). European Union: the European Security Strategy versus national preferences. In E. J. Kichner and J. Sperling, eds., *Global Security Governance: Competing Conceptions of Security in the 21st Century*. Abingdon: Routledge, pp. 113–34.

Kirchner, E. J. and Sperling, J. (2008). *EU Security Governance*. Manchester University Press.

Kjellen, S. (2009). Rapid alerts for crises at the EU level. In S. Olsson, ed., *Crisis Management in the European Union: Cooperation in the Face of Emergencies*. Berlin: Springer, pp. 61–82.

Korski, D. and Gowan, R. (2009). *Can the EU Rebuild Failing States? A Review of Europe's Civilian Capacities*. London: European Council on Foreign Relations.

Krepinevich, A. F. (2009). *7 Deadly Scenarios*. New York: Bantam.

Kuipers, S. L. (2006). *The Crisis Imperative: Crisis Rhetoric and Welfare State Reform in Belgium and the Netherlands*. Amsterdam University Press.

Ladzik, J. (2009). EU military and civilian crisis management operations: the first six years. *European Policy Brief*. London: The Federal Trust for Education and Research, April.

Lagadec, P. (2009). A new cosmology of risks and crises: time for a radical shift in paradigm and practice. *Review of Policy Research*, 26(4), 473–86.

LaPorte, T. R. (2007). Critical infrastructure in the face of a predatory future: preparing for untoward surprise. *Journal of Contingencies and Crisis Management*, 15(1), 60–4.

Larrabee, F. S. and Wilson, P. A. (2012). NATO's shrinking resources. *International Herald Tribune* (May 17), 6.

Larsson, S. (ed.) (2009). *Crisis Management in the European Union: Cooperation in the Face of Emergencies*. Berlin: Springer.

Lavenex, S. (2004). EU external governance in "wider Europe." *Journal of European Public Policy*, 11(4), 680–700.

Lavenex, S. and Wallace, W. (2005). Justice and home affairs: towards a "European public order"? In H. Wallace, W. Wallace, and M. A. Pollack, eds., *Policy-Making in the European Union*. Oxford University Press, pp. 457–80.

Lavenex, S. and Wichmann, N. (2009). The external governance of EU internal security. *Journal of European Integration*, 31(1), 83–102.

Leino, P. (2000). The European Central Bank and legitimacy: Is the ECB a modification of or an exception to the principle of democracy? Harvard Jean Monnet Working Paper 1/01. Cambridge, MA: Harvard Law School.

Leonard, S. (2009). The creation of FRONTEX and the politics of institutionalisation in the European Union External Borders Policy. *Journal of Contemporary European Research*, 5(3), 371–88.

Leonard, S. (2010). EU border security and migration into the European Union: FRONTEX and securitization through practices. *European Security*, 19(2), 231–54.

Lezaun, J. and Groenleer, M. (2006). Food control emergencies and the territorialisation of the European Union. *Journal of European Integration*, 28(5), 437–55.

Liberatore, A. (1999). *The Management of Uncertainty: Learning from Chernobyl*. Amsterdam: Gordon and Breach.

Liddle, R. and Diamond, P. (2010). Aftershock: the coming social crisis in the EU and what is to be done. In E. Marlier and D. Natali, eds., *Europe 2020: Towards a More Social EU?* Oxford: Peter Lang, pp. 69–92.

Lindberg, L. (1965). Decision making and integration in the European community. *International Organization* 19(1), 56–80.

Lindberg, L. N. and Scheingold, S. A. (1970). *Europe's Would-Be Polity: Patterns of Change in the European Community*. Englewood Cliffs, NJ: Prentice-Hall.

Lindblom, C. (1959). The science of "muddling through." *Public Administration Review*, 19(1), 79–99.

Lindborg, C. (2001). The EU rapid reaction force: Europe takes on a new security challenge. Basic Papers: Occasional Papers on International Security Policy, No. 37.

Lindström, G. (2007). *Enter the EU Battlegroups*. Chaillot Paper No. 97. Paris: Institute for Security Studies.

Lindström, G. (2010). Transboundary applications of Common Security and Defence Policy (CSDP). Paper prepared for the Ste Maxime Workshop, April 25–27.

Lomborg, B. (ed.) (2004). *Global Crises, Global Solutions*. Cambridge University Press.

Lundestad, G. (1998). *'Empire' by Invitation: The United States and European Integration*, 1945–1997. Oxford University Press.

Lynch, D. (2009). ESDP and the OSCE. In G. Grevi, D. Helly, and D. Keohane, eds., *European Security and Defence Policy: The First 10 Years (1999–2009)*. Paris: Institute for Security Studies, pp. 139–46.

MacLehose, L., McKee, M., and Weinberg, J. (2002). Responding to the challenge of communicable disease in Europe. *Science*, 295, 2047–50.

Macmaolain, C. (2011). Securing safety, controlling crises: development and misapplication of food law in the European Union. In A. Antoniadis, R. Schutze, and E. Spaventa, eds., *The European Union and Global Emergencies: A Law and Policy Analysis*. Oxford: Hart Publishing, pp. 193–204.

Majone, G. (1996). *Regulating Europe*. Abingdon, UK: Routledge.

Major, C. and Mölling, C. (2010). EU military capabilities – some European troops, but not yet a European army. In E. Greco, N. Pirozzi, and S. Silvestri, eds., *EU Crisis Management: Institutions and Capabilities in the Making*. Rome: Instituto Affari Internazionali, pp. 11–28. NEW: BIB-355

Manners, I. (2002). "Normative power Europe": a contradiction in terms. *Journal of Common Market Studies*, 40(2), 235–58.

Manners, I. (2008). The normative ethics of the European Union. *International Affairs*, 84, 45–60.
March, J. G. and Olsen, J. P. (1984). The new institutionalism: organizational factors in political life. *American Political Science Review*, 78(3), 734–49.
Marples, D. A. (1986). *Chernobyl and Nuclear Power in the USSR*. Basingstoke: Macmillan.
Matzén, N. (2008). EU health security sector: an ascending sector. In A. Boin, M. Ekengren, and M. Rhinard, eds., *Security in Transition: Towards a New Paradigm for the European Union*. Stockholm: National Defence College, pp. 63–77.
Mazower, M. (2002). *The Balkans: A Short History*. London: Modern Library Press.
McInnes, C. (1994). Europe's Jurassic Park: NATO and the end of the Cold War. *Journal of Contingencies and Crisis Management*, 2(1), 21–30.
Mearsheimer, J. J. (1990). Why we will soon miss the Cold War. *The Atlantic* (August), 35–50.
Mearsheimer, J. (1994). The false promise of international institutions. *International Security*, 19(3), 5–49.
Medvedev, Z. A. (1992). *The Legacy of Chernobyl*. New York: W. W. Norton.
Menon, A. (2011). European defence policy from Lisbon to Libya. *Survival*, 53(3), 75–90.
Menon, A. and Sedelmeier, U. (2010). Instruments and intentionality: civilian crisis management and enlargement conditionality in EU security policy. *West European Politics*, 33(1), 75–92.
Mérand, F. (2008). *European Defence Policy: Beyond the Nation State*. Oxford University Press.
Merlingen, M. (2009) EUPM (Bosnia and Herzegovina). In G. Grevi, D. Helly, and D. Keohane, eds., *European Security and Defence Policy: The First 10 Years (1999–2009)*. Paris: Institute for Security Studies, pp. 161–71.
Meunier, S. and Nicolaidis, K. (1999). Who speaks for Europe? The delegation of trade authority in the EU. *Journal of Common Market Studies*, 37(3), 477–501.
Meyer, M. W. and Zucker, L. (1989). *Permanently Failing Organizations*. Thousand Oaks, CA: Sage.
Millstone, E. and Van Zwanenberg, P. (2001). Politics of expert advice: lessons from the early history of the BSE saga. *Science and Public Policy*, 28(2), 99–112.
Missiroli, A. (2003). Ploughshares into swords? Euros for European defence. *European Foreign Affairs Review*, 8(3), 5–33.

Missiroli, A. (Ed.) (2005). *Disasters, Diseases, Disruptions: A New D-drive* for the EU. Chaillot Paper No. 83. Paris: Institute for Security Studies.

Missiroli, A. (2006). Disasters past and present: challenges for the EU. *Journal of European Integration*, 28(5), 423–36.

Missiroli, A. (2008). *Revisiting the European Security Strategy: Beyond 2008*. Brussels: European Policy Centre (May).

Mitsilegas, V., Monar, J., and Rees, W. (2003). *The European Union and Security: Guardian of the People?* London: Palgrave Macmillan.

Monar, J. (2010). Justice and home affairs. *Journal of Common Market Studies*, 48(S1), 143–62.

Moravcsik, A. (1998). *The Choice for Europe: Social Purpose and State Power from Messina to Maastricht*. Ithaca, NY: Cornell University Press.

Myrdal, S. and Rhinard, M. (2010). The European Union's Solidarity Clause: Empty Letter or Effective Tool? UI Occasional Paper No. 2. July.

Nicolaidis, K. (2004). The power of the superpowerless. In T. Lindberg ed., *Beyond Paradise and Power: Europe, America and the Future of a Troubled Partnership*. London: Routledge, pp. 93–120.

Niemann, A. (2006). *Explaining Decisions in the European Union*. Cambridge University Press.

Niemann, A. and Bretherton, C. (2013). EU external policy at the crossroads: the challenge of actorness and effectiveness. *International Relations*, 27(3) (in press).

Nieto, M. and Schinasi, G. J. (2008). *EU Framework for Safeguarding Financial Stability: Towards an Analytical Benchmark for Assessing its Effectiveness*. Documentos Ocasionales No. 0801. Madrid: Banco de Espana.

Nolte, I. M. and Boenigk, S. (2011). Public–nonprofit partnership performance in a disaster context: the case of Haiti. *Public Administration*. 89(4), 1385–402.

Norheim-Martinsen, P. M. (2009). *Matching Ambition with Institutional Innovation: The EU's Comprehensive Approach and Civil-Military Organisation*. Norwegian Defence Research Establishment (FFI), July 3.

North, D. (1990). *Institutions, Institutional Change and Economic Performance*. Cambridge University Press.

Nowak, A. (ed.) (2006a). *Civilian Crisis Management the EU Way*. Chaillot Paper No. 90. Paris: Institute for Security Studies.

Nowak, A. (2006b). Introduction. In A. Nowak, ed., *Civilian Crisis Management the EU Way*. Chaillot Paper No. 90. Paris: Institute for Security Studies, pp. 10–14.

Nowak, A. (2006c). Civilian crisis management within ESDP. In A. Nowak, ed., *Civilian Crisis Management the EU Way*. Chaillot Paper No. 90. Paris: Institute for Security Studies, pp. 15–37.

Nugent, N. (2001). *The European Commission*. London: Palgrave Macmillan.

Nuttall, S. J. (1992). *European Political Cooperation*. Oxford: Clarendon Press.

Nuttall, S. (2005). Coherence and consistency. In C. Hill and M. Smith, eds., *International Relations and the European Union*. Oxford University Press, pp. 91–113.

Nye, J. S., Jr. (1987). Nuclear learning and U.S.–Soviet security regimes. *International Organization*, 41(3), 371–402.

OECD (2003). *Emerging Risks in the 21st Century: An Agenda for Action*. Paris: Organization for Economic Cooperation and Development.

OECD (2011). *Future Global Shocks: Improving Risk Governance*. Paris: Organization for Economic Cooperation and Development.

O'Hanlon, M. and Singer, P. W. (2004). The humanitarian transformation: expanding global intervention capacity. *Survival*, 46(1), 77–100.

Ojanen, H. (2011). The EU as a security actor: in and with the UN and NATO. In S. Blavoukos and D. Bourantonis, eds., *The EU Presence in International Organizations*. London: Routledge, pp. 61–77.

Olsson, E. K. (2005). The dioxin scandal. In S. Larsson, E. K. Olsson, and B. Ramberg, eds., *Crisis Decision Making in the European Union*. Stockholm: CRISMART, pp. 31–64.

Oxford Economics (2010). *The Economic Impact of Air Travel Restrictions due to the Volcanic Ash*. Oxford Economics.

Pastore, F. (2004). Formation, structure and current evolution of the EU entry control system. In N. Walker, ed., *Europe's Area of Freedom, Security, and Justice*. Oxford University Press, pp. 89–142.

Pastore, F. (2007). Europe, migration and development: critical remarks on an emerging policy field. *Development*, 50(4), 56–62.

Patashnik, E. M. (2008). *Reforms at Risk: What Happens after Major Policy Changes Are Enacted*. Princeton University Press.

Pease, K.-K. S. (2012). *International Organizations*. Boston, MA: Longman.

Perrow, C. (2007). *The Next Catastrophe: Reducing Our Vulnerabilities to Natural, Industrial, and Terrorist Disasters*. Princeton University Press.

Peterson, J. (1995). Decision-making in the European Union: towards a framework for analysis. *Journal of European Public Policy*, 2(1), 69–93.

Peterson, J. and Bomberg, E. (1999). *Decision-Making in the European Union*. New York: Macmillan.

Petrov, P. (2010). Early institutionalisation of the ESDP governance arrangements: insights from operations Concordia and Artemis. *European Integration Online Papers (EIoP)*, 14 (Article 8), 1–22.

Phillips, Lord (2000). *The BSE Inquiry: The Report*. London: House of Commons.

Pierre, J. and Peters, B. G. (2009). From a club to a bureaucracy: JAA, EASA, and European aviation regulation. *Journal of European Public Policy*, 16(3), 337–55.

Pierson, P. (1996). The path to European integration: a historical institutionalist analysis. *Comparative Political Studies*, 29(2), 123–63.

Pierson, P. (2001). Increasing returns, path dependence and the study of politics. *American Political Science Review*, 94(2), 251–67.

Pollack, M. (2006). Delegation and discretion in the European Union. In D. Hawkins, D. Lake, D. Nielson, and M. Tierny, M., eds., *Delegation and Agency in International Organizations*. Cambridge University Press, pp. 165–96.

Posner, R. (2004). *Catastrophe: Risk and Response*. New York: Oxford University Press.

Princen, S. and Rhinard, M. (2006). Crashing and creeping: agenda-setting dynamics in the European Union. *Journal of European Public Policy*, 13(7), 1119–32.

Pursiainen, C. (2009). The challenges for European critical infrastructure protection. *Journal of European Integration*, 31(6), 721–39.

Quaglia, L. (2013). Is the European Union governance ready to deal with the next financial crisis? In M. Hallerberg, ed., *Governance Challenges and Innovations: Financial and Fiscal Policy*. Oxford University Press (in press).

Rees, W. (2006). *Transatlantic Counter-Terrorism Cooperation: The New Imperative*. London: Routledge.

Reflection Group on the Future of the EU 2030 (2010). *Project Europe 2030: Challenges and Opportunities*. A report to the European Council by the Reflection Group on the Future of the EU 2030.

Rhinard, M. (2009). European cooperation on future crises: toward a public good? *Review of Policy Research*, 26(4), 439–55.

Rhinard, M. (2010). *Framing Europe: The Policy Shaping Strategies of the European Commission*. Boston, MA: Martinus Nijhoff.

Rhinard, M. and Boin, R. A. (2010). Relief effort reveals some uncomfortable truths. *European Voice* (Jan. 28).

Riggio, D. (2003). EU-NATO cooperation and complementarity between Rapid Reaction Forces. *International Spectator*, 38(3), 47–60.

Rhodes, R. A. W. (1990). Policy networks: a British perspective. *Journal of Theoretical Politics*, 2(3), 292–316.

Rochlin, G. I. (1999). *Trapped in the Net: The Unanticipated Consequences of Computerization*. Princeton University Press.

Rosenthal, U. (1980). *Political Order: Rewards, Punishments and Political Stability*. Alphen a/d Rijn: Sijthoff & Noordhoff.

Rosenthal, U. (1998). Future disasters, future definitions. In E. L. Quarantelli, ed., *What Is a Disaster? Perspectives on the Question*. London: Routledge, pp. 146–59.

Rosenthal, U., Boin, A., and Comfort, L. K. (eds.) (2001). *Managing Crises: Threats, Dilemmas, Opportunities*. Springfield, IL: Charles C. Thomas.

Rosenthal, U., Charles, M., and 't Hart, P. (eds.) (1989). *Coping with Crises: The Management of Disasters, Riots and Terrorism*. Springfield, IL: Charles C. Thomas.

Ruspini, P. (2006). Allargamento Europeo e identità Europea. *Fondazione ISMU, Undicesimo Rapporto sulle migrazioni 2005*. Milan: Franco Angeli, pp. 389–99.

Sabatier, P. and Jenkins-Smith, H. (eds.) (1993). *Policy Change and Learning: An Advocacy Coalition Approach*. Boulder, CO: Westview Press.

Sabel, C. and Zeitlin, J. (2010). *Experimentalist Governance in the European Union: Towards a New Architecture*. Oxford University Press.

Sandholtz, W. (1993) Institutions and collective action: the new telecommunications in Western Europe. *World Politics*, 45(2): 242–70.

Sandholtz W. and Stone Sweet, A. (eds.) (1998). *European Integration and Supranational Governance*. Oxford University Press.

Scannel, D. (2004). Financing ESDP military operations. *European Foreign Affairs Review*, 9(4), 529–49.

Schinasi, G. J. and Teixeira, P. G (2006). The lender of last resort in the European Single Financial Market. IMF working paper 06/127.

Schmitt, C. (1985). *Political Theology*. University of Chicago Press.

Schmitter, P. C. (2000). *How to Democratize the European Union . . . And Why Bother?* London: Rowman & Littlefield.

Schneider, S. K. (1995). *Flirting with Disaster: Public Management in Crisis Situations*. Armonk, NY: M. E. Sharpe.

Seibert, B. H. (2007). African adventure: assessing the European Union's military intervention in Chad and the Central African Republic. MIT Security Studies Program working paper.

Selznick, P. (1957). *Leadership in Administration: A Sociological Interpretation*. Evanston, IL: Row, Peterson.

Simon, L. N. (2010). *Command and Control? Planning for EU Military Operations*. Occasional Paper 81. Paris: European Union Institute for Security Studies.

Simon, L. N. and Mattelaer, A. (2011). *EUnity of Command: The Planning and Conduct of CSDP Operations*. Egmont Paper 41. Brussels: Egmont Institute.

Smith, C. (2011). EU–NATO relations. Work Package 4, Politics and Ideology. Nordregio, Sweden: EuroBroadMap.

Smith, M. (2003). The framing of European foreign and security policy: towards a post-modern policy framework? *Journal of European Policy*, 10(4), 556–75.

Sperling, J. (2010). The European Union and NATO: subordinate partner, cooperative pillar, compering pole? In S. Blaroukos and D. Bourantonis, eds., *The EU Presence in International Organizations*. London: Routledge, pp. 33–60.

Staelgrave, S. and 't Hart, P. (2008). Dutroux and dioxin: crisis investigations, elite accountability and institutional reform in Belgium. In A. Boin, A. McConnell, and P. 't Hart, eds., *Governing after Crisis: The Politics of Investigation, Accountability and Learning*. Cambridge University Press, pp. 148–81.

Stern, E. and Sundelius, B. (2002). Crisis management in Europe: an integrated regional research and training program. *International Studies Perspective*, 3(1), 71–88.

Stone Sweet, A., Fligstein, N., and Sandholtz, W. (2001). The institutionalization of European space. In A. Stone Sweet, W. Sandholtz, and N. Fligstein, eds., *The Institutionalization of Europe*. Oxford University Press.

Suchman, M. C. (1995). Managing legitimacy: strategic and institutional approaches. *Academy of Management Review*, 20(3), 571–610.

Sundelius, B. (2005). Disruptions: functional security for the EU. In A. Missiroli, ed., *Disasters, Diseases, Disruptions: A New D-Drive for the European Union*. Chaillot Paper No. 83. Paris: Institute for Security Studies, European Union, pp. 67–84.

Sundelius, B. and Grönvall, J. (2004). Strategic dilemmas of biosecurity in the European Union. *Biosecurity and Bioterrorism*, 2(1), 17–23.

Sylves, R. (2008). *Disaster Policy and Politics: Emergency Management and Homeland Security*. Washington, DC: CQ Press.

Taleb, N.N. (2007). *The Black Swan: The Impact of the Highly Improbable*. New York: Random House.

Taylor, P. (2007). *The End of European Integration: Anti-Europeanism Examined*. Abingdon, UK: Routledge.

Tenner, E. (1995). *Why Things Bite Back*. New York: Vintage.

Thwaites, N. (2006). Eurojust: beacon in EU judicial co-operation. *International Review of Penal Law*, 77, 293–8.

Turner, B. A. (1978). *Man-made Disasters*. London: Wykeham.

Uçarer, E. M. (2002). Guarding the borders of the European Union: paths, portals, and prerogatives. In S. Lavenex and E. M. Uçarer, eds., *Migration and the Externalities of European Integration*. Lanham, MD: Lexington Books, pp. 15–32.

Ulriksen, S., Gourlay, C., and Mace, C. (2004). Operation Artemis: the shape of things to come? *International Peacekeeping*, 11(3), 508–25.

US NRF (2008). *National Response Framework*. Washington, DC: Department of Homeland Security.

Vahl, R. (1997). *Leadership in Disguise: The Role of the European Commission in EC Decision-Making on Agriculture in the Uruguay Round*. Aldershot, UK: Ashgate.

Valasek, T. (2007). The roadmap to better EU–NATO relations. *Centre of European Reform Briefing Note*. London: Centre for European Reform.

Van Den Wyngaert, C. (2006). Eurojust and the European Public Prosecutor in the *Corpus Juris* model: water and fire? In N. Walker, ed., *Europe's Area of Freedom, Justice and Security*. Oxford University Press, pp. 201–39.

Van Ham, P. and Medvedev, S. (2002). *Mapping European Security after Kosovo*. Manchester University Press.

Varwick, J. and Koops, J. (2009). The European Union and NATO: 'shrewd interorganizationalism' in the making? In K. E. Jorgensen, ed., *The European Union and International Organizations*. London and New York: Routledge, pp. 101–30.

Vincent, K. (2004). 'Mad cows' and Eurocrats: Community responses to the BSE crisis. *European Law Journal*, 10(5), 499–517.

Von Hippel, K. (2005). *Europe Confronts Terrorism*. New York: Palgrave.

Waever, O., Buzan, B., Kelstrup, M., and Lemaitre, P. (1993). *Identity, Migration, and the New Security Order in Europe*. London: Pinter.

Wagner, W. (2003). Why the EU's Common Foreign and Security Policy will remain intergovernmental: a rationalist institutional choice analysis of European crisis management policy. *Journal of European Public Policy*, 10(4), 576–95.

Walker, N. (eds.) (2004). *Europe's Area of Freedom, Security and Justice*. Oxford University Press.

Wall, S. (2008). *A Stranger in Europe: Britain and the EU from Thatcher to Blair*. Oxford University Press.

Walsh, J. I. (2006). Intelligence sharing in the European Union: institutions are not enough. *Journal of Common Market Studies*, 44(3), 625–43.

Waltz, K. (1979). *Theory of International Politics*. London: Addison-Wesley.

Waltz, K. (2000). Structural realism after the Cold War. *International Security*, 25(1), 5–41.

Waltz, K. N. (1993). The emerging structure of international politics. *International Security*, 18(2), 44–79.
Wendling, C. (2010). Explaining the emergence of different European Union crisis and emergency management structures. *Journal of Contingencies and Crisis Management*, 18(2), 74–82.
Wessels, W. and Bopp, F. (2008). The institutional architecture of CFSP after the Lisbon Treaty: constitutional breakthrough or challenges ahead? Research Paper No. 10. June. Challenge (Sixth EU Framework Programme).
Wettig, J. and Porter, S. (1998). Seveso Directive: background, contents and requirements. In C. Kirchsteiger, ed., *Risk Assessment and Management in the Context of the Seveso II Directive*. Amsterdam: Elsevier, pp. 27–46.
White, E. (2011). The constitutional apparatus of the European Union: the perspective of the European Commission. In A. Antoniadus, R. Schutze, and E. Spaventa, eds., *The European Union and Global Emergencies: A Law and Policy Analysis*. Oxford: Hart Publishing, pp. 261–72.
Wilson, J. Q. (1989). *Bureaucracy: What Government Agencies Do and Why They Do It*. New York: Basic Books.
Wolf, M. (2012). The Union outlived de Tocqueville. *Financial Times* (April 18), 11.
Wright, R. and Auvinen, J. (2009). What ambitions for EU civilian crisis management? In A. De Vasconcelos, ed., *What Ambitions for European Defence in 2020?*. Paris: European Union Institute for Security Studies, pp. 117–28.
Zeitlin, J. (2011). *Transnational Transformations of Governance: The European Union and Beyond*. Amsterdam: Vossiuspers.
Zimmermann, D. (2006). The European Union and post-9/11 counterterrorism: a reappraisal. *Studies in Conflict and Terrorism*, 29(1), 123–45.
Zürn, M. and Checkel, J. T. (2005). Getting socialized to build bridges: constructivism and rationalism, Europe and the nation-state. *International Organization*, 59, 1045–79.

Index